SpringerBriefs in Computer Science

Series Editors

Stan Zdonik
Peng Ning
Shashi Shekhar
Jonathan Katz
Xindong Wu
Lakhmi C. Jain
David Padua
Xuemin Shen
Borko Furht
V. S. Subrahmanian
Martial Hebert
Katsushi Ikeuchi
Bruno Siciliano

For further volumes:
http://www.springer.com/series/10028

Muhammad Ismail · Weihua Zhuang

Cooperative Networking in a Heterogeneous Wireless Medium

 Springer

Muhammad Ismail
Weihua Zhuang
Department of Electrical and Computer Engineering
University of Waterloo
Waterloo, ON
Canada

ISSN 2191-5768   ISSN 2191-5776 (electronic)
ISBN 978-1-4614-7078-6   ISBN 978-1-4614-7079-3 (eBook)
DOI 10.1007/978-1-4614-7079-3
Springer New York Heidelberg Dordrecht London

Library of Congress Control Number: 2013933595

Printed on acid-free paper

Springer is part of Springer Science+Business Media (www.springer.com)

Preface

The past decade has witnessed an increasing demand for wireless communication services, which have extended beyond telephony services to include video streaming and data applications. This results in a rapid evolution and deployment of wireless networks, including the cellular networks, the IEEE 802.11 wireless local area networks (WLANs), and the IEEE 802.16 wireless metropolitan area networks (WMANs). With overlapped coverage from these networks, the wireless communication medium has become a heterogeneous environment with a variety of wireless access options. Currently, mobile terminals (MTs) are equipped with multiple radio interfaces in order to make use of the available wireless access networks. In such a networking environment, cooperative radio resource management among different networks will lead to better service quality to mobile users and enhanced performance for the networks.

In this brief, we discuss decentralized implementation of cooperative radio resource allocation in a heterogeneous wireless access medium for two service types, namely single-network and multi-homing services. In Chap. 1, we first give an overview of the concept of cooperation in wireless communication networks and then we focus our discussion on cooperative networking in a heterogeneous wireless access medium through single-network and multi-homing services. In Chap. 2, we present a decentralized optimal resource allocation (DORA) algorithm to support MTs with multi-homing service. The DORA algorithm is limited to a static system model, without new arrival and departure of calls in different service areas, with the objective of identifying the role of each entity in the heterogeneous wireless access medium in such a decentralized architecture. In Chap. 3, we discuss the challenges that face the DORA algorithm in a dynamic system and present a sub-optimal decentralized resource allocation (PBRA) algorithm that can address these challenges. The PBRA algorithm relies on short-term call traffic load prediction and network cooperation to perform the decentralized resource allocation in an efficient manner. We present two design parameters for the PBRA algorithm that can be properly chosen to strike a balance between the desired performance in terms of the allocated resources per call and the call blocking probability, and between the performance and the implementation complexity. In Chap. 4, we further extend the radio resource allocation problem to consider the simultaneous presence of both single-network and multi-homing services in the networking environment. We first develop a centralized optimal resource allocation (CORA) algorithm to find the optimal network selection for MTs with single-network service and the corresponding optimal bandwidth allocation for MTs with single-network and multi-homing services. Then we present a decentralized implementation for the radio resource allocation using a decentralized sub-optimal resource allocation (DSRA) algorithm. The DSRA algorithm gives the MTs an active role in the resource allocation operation, such that an MT with single-network service can select the best available network at its location and asks for its required bandwidth, while an MT with multi-homing service can determine the required bandwidth share from each network in order to satisfy its total required bandwidth. Finally, we draw conclusions and outline future research directions in Chap. 5.

January 2013   Muhammad Ismail
                                                                               Weihua Zhuang

Contents

Chapter 1
Introduction

Cooperation in wireless communication networks is expected to play a key role in addressing performance challenges of future wireless networks. Hence, both academia and industry have issued various proposals to employ cooperation so as to improve the wireless channel reliability, increase the system throughput, or achieve seamless service provision. In the existing proposals, cooperation comes at three different levels, namely among different users, among users and networks, and among different networks. In fact, the current nature of the wireless communication medium constitutes the driving force that motivates the last cooperation level, i.e. cooperation among different networks. Currently, the wireless communication medium is a heterogeneous environment with various wireless access options and overlapped coverage from different networks. Cooperation among these different networks can help to improve the service quality to mobile users and enhance the performance for the networks. In this chapter, we first introduce the concept of cooperation in wireless communication medium, then focus on cooperative networking in heterogeneous wireless access networks and its potential benefits for radio resource management.

1.1 Cooperation in Wireless Communication Networks

According to Oxford dictionary, cooperation is defined as *"the action or process of working together to the same end"*, which is the opposite of working separately (selfishly) in competition. Over the years, this concept has been studied in social sciences and economics in order to maximize the individuals' profit. Only recently, cooperation has been introduced to wireless communications as a promising response to the challenges that face the development of the wireless networks, which include the scarcity of radio spectrum and energy resources and necessity to provide adequate user mobility support.

Regardless of the networking environment, three cooperation scenarios can be distinguished based on various studies in literature [71]. The first scenario employs cooperation among different entities to improve the wireless communication channel

M. Ismail and W. Zhuang, *Cooperative Networking in a Heterogeneous Wireless Medium*, SpringerBriefs in Computer Science, DOI: 10.1007/978-1-4614-7079-3_1, © The Author(s) 2013

reliability through spatial diversity and data relaying [31, 48]. The second scenario employs cooperation to improve the achieved throughput via aggregating the offered resources from different cooperating entities [24, 27, 28, 68]. Finally, cooperation is used to guarantee service continuity and achieve seamless service provision [16, 33, 62, 63]. These cooperation scenarios are explained in more details in the following.

1.1.1 Cooperation to Improve Channel Reliability

The wireless communication medium is challenged by several phenomena that decrease its reliability, including path loss, shadowing, fading, and interference. Cooperation in wireless communication networks can improve the communications reliability against these impairments.

First, cooperation can mitigate the wireless channel fading through cooperative spatial diversity [31, 48]. Specifically, when the direct link between the source and destination nodes is unreliable, other network entities can cooperate with the source node and form a virtual antenna array to forward data towards the destination. Through the virtual antenna array, different transmission paths with independent channel coefficients exist between the source and destination nodes. Hence, the destination node receives several copies of the same transmitted signal over independent channels. Using the resulting spatial diversity, the destination node combines the received signals from the cooperating entities in detection in order to improve the

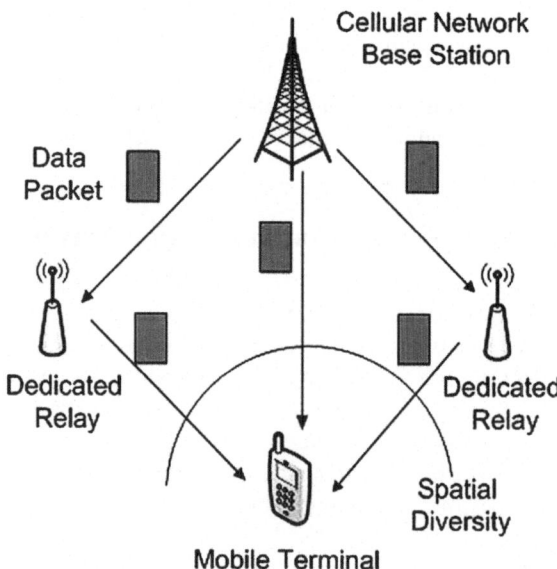

Fig. 1.1 Cooperative spatial diversity

transmission accuracy. Cooperative spatial diversity is illustrated in Fig. 1.1 for a downlink transmission from a base station (BS) to a mobile terminal (MT). In this figure, the BS transmits its data packets towards the MT using the help of dedicated relays that create a virtual antenna array. This concept has proven to be very useful to improve transmission accuracy for situations where it is infeasible to employ multiple transmission and reception antennas at different nodes for traditional spatial diversity. In cooperative spatial diversity, a cooperating entity is simply a relay node with an improved channel condition over the direct source-destination channel. The relay node can be either an MT or a dedicated relay as in Fig. 1.1.

In addition, cooperation can help to reduce the resulting interference due to the broadcast nature of the wireless communication medium. In general, the resulting interference reduces the signal-to-interference-plus-noise ratio (SINR) at the receiving nodes which degrades the detection performance. Through cooperative relays, the transmitted power from the original source node can be significantly reduced due to the better channel conditions of the relaying links. This can greatly reduce the interference region [70], which is illustrated in Fig. 1.2. Finally, cooperation can solve the hidden terminal problem, which also results in interference reduction and improves channel reliability [2].

1.1.2 Cooperation to Improve the Achieved Throughput

An enhanced channel reliability through cooperative spatial diversity and relaying directly results in an improved achieved throughput. In addition, cooperation can help

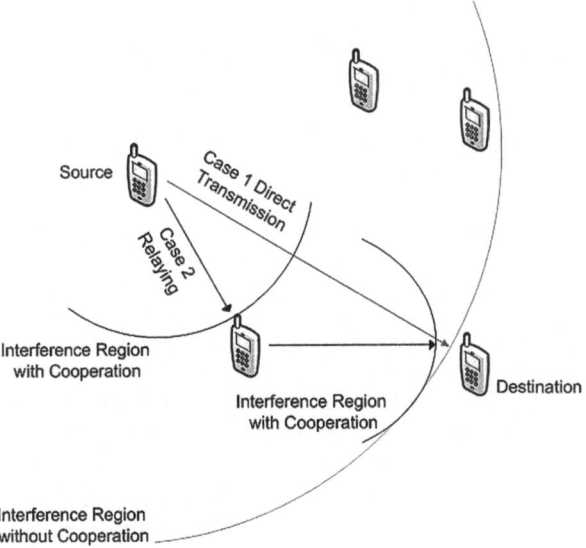

Fig. 1.2 Cooperative interference reduction

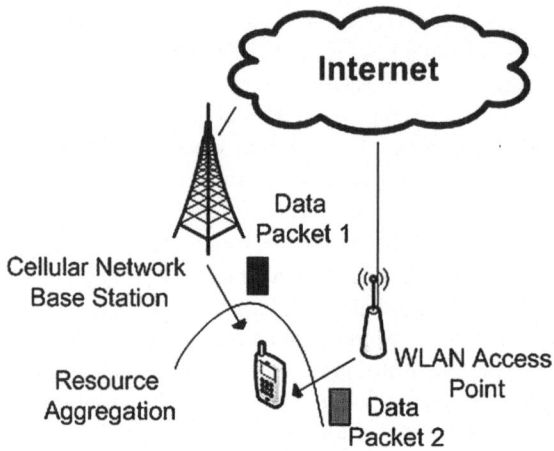

Fig. 1.3 Cooperative resource aggregation

to improve the achieved throughput via aggregating the offered resources from different cooperating entities [24, 27, 28, 68]. Unlike cooperative spatial diversity strategies which take place at the physical layer [31, 48], cooperative resource aggregation strategies take place at the network layer [24, 27, 28] and transport layer [72]. In this scenario, data packets are transmitted from a source to the destination through multiple paths. However, unlike cooperative spatial diversity, the transmitted data packets through different paths are not copies of the same transmitted signal. Instead, different data packets are transmitted through these paths. This results in an increase in the total transmission data rate between the source and destination nodes. This concept is illustrated in Fig. 1.3, where the resources from cooperating cellular network BS and wireless local area network (WLAN) access point (AP) are aggregated in order to support a high data rate for the MT. In cooperative resource aggregation, the cooperating entities can be MTs, BSs, or APs with sufficient resources (e.g. bandwidth), such that when aggregated, the total transmission data rate from the source to the destination can be increased.

1.1.3 Cooperation to Support Seamless Service Provision

In communication networks, call blocking refers to a new call that is not allowed to enter service due to resource unavailability, while call dropping refers to a call that is forced to terminate prematurely [23]. In general, mobile users are more sensitive to call dropping than call blocking. Depending on the networking environment, call dropping may interrupt service continuity for different reasons (e.g. in cellular networks this can be due to user mobility among cells, while in cognitive radio networks this can be due to the primary user activities). Employing cooperative strategies at

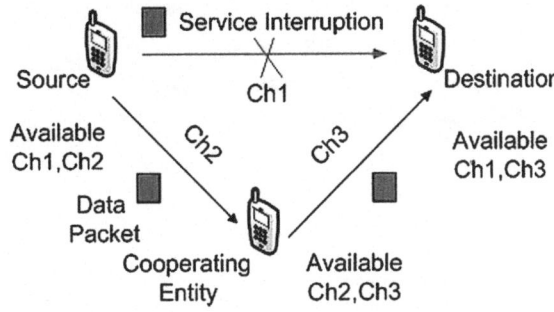

Fig. 1.4 Cooperative seamless service provision

link, network, and transport layers can better guarantee service continuity for ongoing calls [16, 33, 62, 63]. In cooperative seamless service provision, when service is interrupted along the direct link from the source to the destination, cooperating entities can help to create an alternative path in order to support service continuity. This concept is illustrated in Fig. 1.4, where service is interrupted along the direct link between the source and destination nodes (Ch1), yet it still can be continued using another cooperative path (Ch2, Ch3). In cooperative seamless service provision, a cooperating entity can be an MT, BS, or AP which can create a substitute path between the source and destination nodes.

All three cooperation scenarios (cooperative spatial diversity, cooperative resource aggregation, and cooperative seamless service provision) can occur in different networking environments which include cellular networks, cognitive radio networks, mobile ad hoc networks, vehicular ad hoc networks, etc. [4, 7, 17, 35, 36, 56, 71]. In these scenarios, cooperation can take place at different levels, which can be among mobile users, among mobile users and networks, and among different networks [71]. Currently, the wireless communication medium is a heterogeneous environment with overlapped coverage from different networks [28]. Such an environment motivates the third cooperation level, i.e. cooperation among different networks. Cooperative networking can be beneficial for both mobile users and network operators [26]. In the following, we first present the heterogeneous wireless access medium, then discuss the potential benefits of cooperative networking in this environment.

1.2 The Heterogeneous Wireless Access Medium

Currently, there exist several wireless networks that offer a variety of access options, such as the cellular networks, the IEEE 802.11 WLANs, the IEEE 802.16 wireless metropolitan area networks (WMANs), etc. These different wireless networks have complimentary service capabilities. For instance, while the IEEE 802.11 WLANs can support high data rate services in hot spot areas, the cellular networks and the IEEE 802.16 WMANs can provide broadband wireless access over long distances

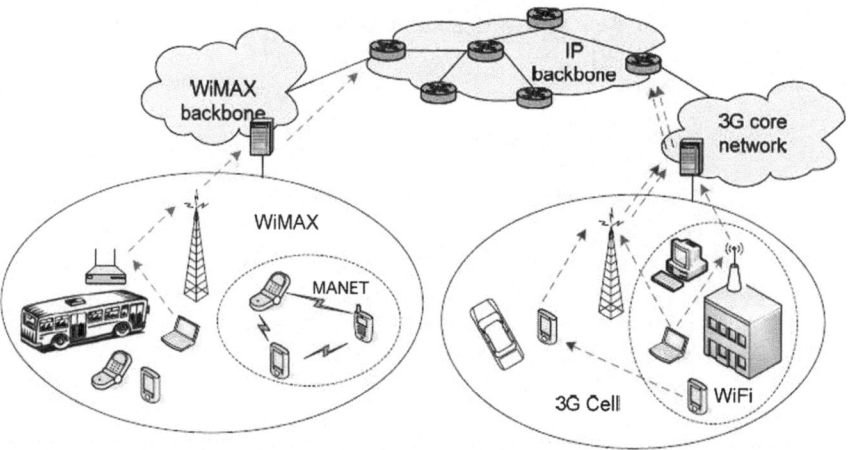

Fig. 1.5 An illustration of heterogeneous wireless communication network architecture

and serve as a backbone for hot spots [26]. As a result, these networks will continue to coexist. This turns the wireless communication medium into a heterogeneous environment with overlapped coverage from different networks.

1.2.1 The Network Architecture

The basic components of the heterogeneous wireless communication network architecture are MTs, BSs/APs, and a core Internet protocol (IP) based network [12], as shown in Fig. 1.5.

Currently, mobile users are viewed as service recipients in the network operation, with passive transceivers that operate under the control of BSs or APs. It is envisioned that future MTs will be more powerful and play a more active role in the network operation and service delivery. Currently, some MTs are equipped with multiple radio interfaces in order to make use of the available access opportunities in this networking environment. Moreover, an MT is able to maintain multiple simultaneous associations with different radio access networks using the multi-homing capabilities. Fixed network components, such as BSs and APs, provide a variety of services to MTs. These services include access to the Internet and mobility and resource management. Finally, the core network serves as the backbone network with Internet connectivity and packet data services.

1.2.2 Potential Benefits of Cooperative Networking

Despite the fierce competition in the wireless service market, the aforementioned wireless networks will coexist due to their complementary service capabilities. In this heterogeneous wireless access medium with overlapped coverage from different networks, cooperative networking will lead to better service quality to mobile users and enhanced performance for the networks.

As for mobile users, cooperative networking solutions for heterogeneous wireless networks can result in two major advantages. The first advantage is that mobile users can enjoy an always best connection. This means that a mobile user can always be connected to the best wireless access network available at his/her location. Traditionally, an MT can keep its connection active when it moves from one attachment point to another through handoff management [3]. Hence, mobile users can enjoy an always connected experience. This is enabled by horizontal handoff, which represents a handoff within the same wireless access network, as in the handoff between two APs in a WLAN or between two BSs in a cellular network. However, in the presence of various wireless access networks with overlapped coverage, the user experience is now shifted from always connected to always best connected. The always best connected experience is mainly supported by vertical handoffs among different networks. A vertical handoff represents a handoff between different wireless access networks, as in the handoff between a BS of a cellular network and an AP of a WLAN. Unlike horizontal handoffs, vertical handoffs can be initiated for convenience rather than connectivity reasons. Hence, vertical handoffs can be based on service cost, coverage, transmission rate, quality-of-service (QoS), information security, and user preference. Through cooperative networking, the inter-network vertical handoffs can be provided in a seamless and fast manner. This can support a reliable end-to-end connection at the transport layer, which preserves service continuity and minimizes disruption. Hence, this represents a cooperative seamless service provision scenario. The second advantage of cooperative networking for mobile users is that users can enjoy applications with high required data rates through aggregating the offered radio resources from different networks. This is enabled by the multi-homing capabilities of MTs, where users can receive their required radio resources through different networks and use multiple threads at the application layer. In this context, cooperation is required among different networks so as to coordinate their allocated radio resources to the MTs such that the total resource allocation from multiple networks satisfies the user total required data rate. Hence, this falls under the category of cooperative resource aggregation.

In addition, service providers can benefit from cooperative networking to enhance network performance in many ways. For instance, multiple heterogeneous networks can cooperate to provide a multi-hop backhaul connection in a relay manner. This results in an increase in these networks coverage area at a reduced cost as compared to deploying more BSs for coverage extension. Also, load balancing among different networks can be supported through cooperative networking which helps in avoiding call traffic overload situations. Moreover, cooperative networking can achieve energy saving for green radio communications. Networks with overlapped coverage area can alternately switch their BSs on and off according to spatial and temporal fluctuations in call traffic load, which reduces their energy consumption and provides an acceptable QoS performance for the users [26].

In this brief, we mainly focus on cooperation among different networks in a heterogeneous wireless access medium to enhance the mobile users perceived QoS through radio resource management mechanisms. Specifically, we will adopt the cooperative resource aggregation and cooperative seamless service provision concepts for radio resource allocation to provide an improved service quality for mobile users. Hence, in the following, we first present a literature survey on radio resource allocation mechanisms in a heterogeneous wireless access medium.

1.3 Radio Resource Allocation in Heterogeneous Wireless Access Medium

Radio resource allocation mechanisms aim to efficiently utilize the available resources to satisfy QoS requirements of different users. Different types of services impose different requirements in terms of resource allocation. In general, two types of services can be distinguished.

1. Inelastic calls, which require a fixed resource allocation that is available during the connection duration. This is similar to the constant bit rate (CBR) service class in asynchronous transfer mode (ATM) networks. One example of this class is the traditional voice telephony.
2. Elastic calls, which can adapt their required resources according to the network's instantaneous call traffic load. A minimum resource allocation is required in order to satisfy a minimum service quality. However, more resources can be allocated up to a maximum value to improve data delivery performance of the end-to-end connection. Hence, this class is similar to the variable bit rate (VBR) service class in ATM networks. Two examples of this service class are video and data calls. The key difference between video and data calls is the impact of the allocated resources on the call presence in the system [37]. For video calls, the amount of the allocated resources influences the perceived video quality experienced on the video terminal, while it does not affect the video call duration. On the other hand, the resource allocation to a data call affects its throughput and thus its duration.

Currently, there exist different wireless access networks with different service capabilities in terms of bandwidth, coverage area, cost, and so on. The available resources from these networks can be used to satisfy the QoS requirements for different service types. However, this utilization should be performed in an efficient way. Hence, a resource allocation framework that can satisfy the QoS requirements of different connections while making efficient utilization of available resources is needed. This framework is presented in the following sub-section.

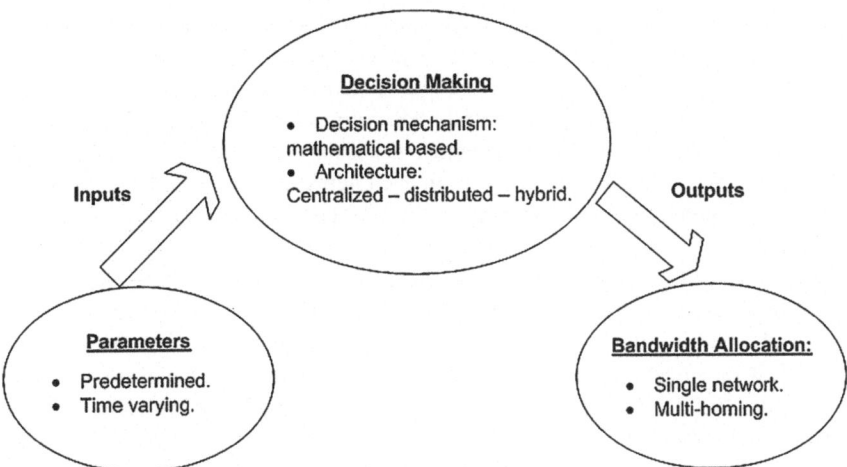

Fig. 1.6 Resource allocation framework

1.3.1 Radio Resource Allocation Framework

The resource allocation problem in a heterogeneous wireless access environment can be viewed as a decision making process [52]. This can be represented by the framework shown in Fig. 1.6. The framework has three components, namely, inputs, decision making, and outputs, as discussed in the following [52].

- **Inputs**
  In order to determine an optimal resource allocation for a given connection in a heterogeneous wireless access medium, a set of information needs to be gathered. This set of information is used as inputs to the decision making engine. These inputs can be divided into two categories. One category includes predetermined inputs, which are set a priori and remain unchanged for the connection duration. They include the user preferences such as cost, security, and power consumption. Also, they include the application type along with its QoS constraints such as required bandwidth. The other category includes the time varying inputs. These vary during the connection duration and are monitored continuously. They include network call traffic load, the available radio coverage, and the connection holding time.
- **Decision Making**
  With all gathered information, resource allocation schemes deploy various decision making techniques in order to reach the best possible allocation. The decision making process should define a decision mechanism and a decision place. The decision mechanism provides a means for determining the optimal resource allocation. In general, the decision mechanism employs a profit/utility function in order to assess the resulting users' satisfaction from the allocated resources. Decision mechanisms can employ stochastic programming, game theory, or convex

optimization to determine the optimal allocation. Another important factor in the decision making process is the decision place. In literature, three types of architectures can be defined, namely centralized, distributed, and hybrid architectures. In a centralized architecture a central node, with a global view of all resources of different networks and service demands, makes the decision, while in a distributed approach the decision is made either in each network or eventually in the MT. A hybrid architecture is a mix of both centralized and distributed approaches.

- **Outputs**

  In literature, the resource allocation mechanisms in a heterogeneous wireless access medium can be divided in two categories [25]. The first category utilizes a single interface of an MT, so that the MT obtains its required resources from a single access network (which is the best available network at the user's location). We refer to this category as single-network resource allocation mechanisms. In single-network resource allocation mechanisms, the objective is to find the optimal network assignment for different users (i.e. which user is assigned to which network) and the optimal resource allocation within this network based on some predefined criteria. As a result, the output of the decision making process is the network assignment and the amount of resources allocated from the network. The second category of the resource allocation mechanisms utilizes multiple radio interfaces of an MT simultaneously to support the service requirement. We refer to this category as multi-homing resource allocation mechanisms. The MT in this type of solutions obtains its required resources from all available wireless access networks. Hence, in this category the decision making process output is the amount of resources allocated from various networks to a given connection.

Table 1.1 Single-network resource allocation mechanisms in a heterogeneous wireless medium

Reference	Mechanism	Objective	Architecture
[13]	Stochastic programming	To maximize the allocations under demand uncertainty while minimizing underutilization of different networks and users' rejection	Centralized
[64]	Convex optimization	To maximize the minimum throughput among all users in the heterogeneous networks	Centralized
[51]	Convex optimization	To maximize the total welfare of each network, with the aim of satisfying the signal quality requirements of all mobile users in a CDMA cellular network and controlling the optimum collision probability in a WLAN	Centralized
[8]	Convex optimization	To find close to optimum allocation for a given set of voice users with minimum QoS requirements and a set of best effort users	Distributed

1.3.2 Radio Resource Allocation Mechanisms

In this sub-section, radio resource allocation mechanisms from single-network and multi-homing solutions are reviewed. The different mechanisms are discussed in terms of their objectives and the decision making architectures. We start with the single-network mechanisms, then present the multi-homing mechanisms.

Single-Network Radio Resource Allocation Mechanisms

Table 1.1 summarizes some mechanisms employed in the single-network resource allocation. For the mechanisms with a centralized architecture, a central controller is assumed to select the best network for a given connection from a set of available wireless networks, and then performs the resource allocation for that connection from the selected network. For the distributed mechanism in Table 1.1, an MT selects the best available network and the selected network then performs the resource allocation for the connection. In general, the selection of the best available network depends on a predefined criterion [29]. One criterion is the received signal strength (RSS) [41], where the MT is assigned to the wireless network with the highest RSS from its BS or AP among all available networks. Another network selection criterion is the offered bandwidth [58], where the MT is assigned to the network BS/AP with the largest offered bandwidth. Moreover, different network selection criteria, such as RSS, offered bandwidth, and monetary cost, can be combined in a utility function and the MT network assignment is based on the results of this function associated with the BSs/APs of the candidate networks [43]. The single-network resource allocation mechanisms suffer from a limitation that an incoming call is blocked if no network in the service area can individually satisfy the call required QoS. As a result, these mechanisms do not fully exploit the available resources from different networks.

Multi-homing Radio Resource Allocation Mechanisms

In multi-homing solutions for resource allocation, each MT can obtain its required resources for a specific application from all available wireless access networks. This has the following advantages [14]:

1. With multi-homing capabilities, the available resources from different wireless access networks can be aggregated to support applications with high required data rates (e.g. video streaming and data calls) using multiple threads at the application layer;
2. Multi-homing solutions allow for better mobility support, since at least one of the MT radio interfaces will remain active, at a time, during the call duration;
3. The multi-homing concept can reduce the call blocking rate and improve the overall system capacity.

Some mechanisms for multi-homing resource allocation are summarized in Table 1.2. All the centralized mechanisms assume the existence of a central resource manager that determines the optimum resource allocation from each available network to satisfy the MT required QoS.

Table 1.2 Multi-homing resource allocation mechanisms in a heterogeneous wireless medium

Reference	Mechanism	Objective	Architecture
[46, 66]	Cooperative game	To form a coalition among different available wireless access networks to offer bandwidth to a new connection	Centralized
[45, 47]	Non cooperative game	To develop a profit oriented bandwidth allocation mechanism (The requested bandwidth is allocated to a new connection from all available networks based on the available bandwidth in each network. All networks compete with each other to maximize their profit.)	Centralized
[39]	Utility function based	To allocate bandwidth to both CBR and VBR connections from all available networks depending on utility fairness for each type of service	Centralized

1.3.3 Cooperative Radio Resource Allocation

Almost all the existing research works in literature on radio resource allocation in a heterogeneous wireless access medium focus on supporting either a single-network or a multi-homing service. However, it is envisioned that both service types will coexist in the future networks [27, 29]. This is because not all MTs are equipped with multi-homing capabilities, and not all services require high resource allocation that calls for a multi-homing support. As a result, some MTs will have to utilize a single-network service. Moreover, even for an MT with a multi-homing capability, the MT utilization of the multi-homing service should depend on its residual energy. Hence, when no sufficient energy is available at the MT, the MT should switch from a multi-homing service to a single-network one where the radio interface of the best available wireless network is kept active while all other interfaces are switched off to save energy. This motivates the requirement to develop a radio resource allocation mechanism that can support both single-network and multi-homing services in a heterogeneous wireless access medium. However, there are many technical challenges, as discussed in the following.

Decentralized Implementation

From the literature survey summarized in Tables 1.1 and 1.2, it is clear that, except for the work in [8], almost all radio resource allocation mechanisms need a central resource manager in order to meet service quality requirements in such a heterogeneous wireless access medium. In addition, the work in [8] is to support MTs with only single-network service. The need for the central resource manager for single-network services is due to the fact that a global view over the individual networks' status is required in order to select the best available wireless access network given the MT required QoS. For multi-homing services, the central resource manager coordinates the allocated resources from different networks such that the total resource allocation to a given MT equals to the total required resources by this MT. Hence, the central resource manager should have a global view over network available resources, and perform network selection for MTs with single-network services and resource allocation for MTs with single-network and multi-homing services. However, the assumption of the presence of this central resource manager is not practical in a case that the networks are operated by different service providers. This is because the central resource manager would raise some issues [28]:

1. The central resource manager is a single point of failure. Hence, if it breaks down, the whole single-network and multi-homing services fail and this may extend to the operation of the different networks;
2. Which network should be in charge of the operation and maintenance of this central resource manager, taking account that the network in charge will control the radio resources of other networks;
3. Modifications are required in different network structures in order to account for this central resource manager.

As a result, it is desirable to have a decentralized implementation of the radio resource allocation. In this context, an MT with single-network service can select the best wireless access network available at its location and asks for its required resources from this network. While an MT with multi-homing service can ask for the required resources from each available network so as to satisfy its total required service quality. Each network then can perform its own resource allocation and admission control without the need for a central resource manager. However, with users and service requests following stochastic mobility and traffic models, achieving the optimal allocation for a given connection at any point of time would trigger reallocations of a whole set of connections. This will take place with every service request arrival or departure and a considerable amount of signalling information has to be exchanged among different network entities. Hence, through network cooperation, we aim to develop an efficient decentralized implementation of the radio resource allocation that balances the resource allocation with the associated signalling overhead. Through cooperative resource allocation, different networks can coordinate their resource allocation in order to support the QoS of each call, satisfy a target call blocking probability, and eliminate the need for a central resource manager while reducing the amount of signalling overhead over the air interface.

Service Differentiation

In general, mobile users are the subscribers of different networks. As a result, the service requests of different MTs should not be treated in the same manner by each network. Instead, it is more practical that each network gives a higher priority in allocating its resources to its own subscribers as compared to other users. Hence, a priority mechanism should be in place to enable each network to assign different priorities to MTs on its resources.

Considering the aforementioned challenges in designing a resource allocation mechanism to support both single-network and multi-homing services in a dynamic environment, we will take the following steps to present radio resource allocation solutions.

1. Static multi-homing radio resource allocation in Chap. 2: In this step, we will investigate a system model with only multi-homing calls, and without considering the arrival of new calls or departure of existing ones. This simplifies the problem under consideration due to two reasons. Firstly, in the absence of a network assignment problem we focus on finding the optimal resource allocation from each network to a given connection in order to satisfy its total required bandwidth. Secondly, due to the static nature of the system model, there are no perturbations associated with the number of MTs in the system. Hence, no resource reallocations are necessary, and the signalling between MTs and BSs/APs occur only in the call setup. We aim to develop a decentralized implementation of the radio resource allocation and identify the role of each network entity in this architecture. In addition, we shall enable each network to give a higher priority in allocating its resources to its own subscribers as compared to other users;
2. Dynamic multi-homing radio resource allocation in Chap. 3: We consider the stochastic mobility and traffic models for the users and service requests. The system experiences perturbations in the call traffic load. This triggers resource reallocations for all the existing connections, and results in a considerable amount of signalling overhead. Hence, we will extend the resource allocation in step 1 in order to provide an efficient radio resource allocation mechanism that can balance the resource allocation with the associated signalling overhead through short-term call traffic load prediction and network cooperation;
3. Single-network and multi-homing radio resource allocation mechanism in Chap. 4: We extend the ideas presented in Chaps. 2 and 3 to include single-network calls in the system model. Hence, the radio resource allocation mechanism is of twofold: to determine the network assignment of MTs with single-network service to the available wireless access networks, and to find the corresponding resource allocation to MTs with single-network and multi-homing services. The framework gives an active role to the MTs in the resource allocation operation through network selection and resource requests.

1.4 Summary

In this chapter, three cooperation scenarios are discussed, namely cooperative spatial diversity, cooperative resource aggregation, and cooperative seamless service provision. The cooperation scenarios can take place at three different levels, which are among users, between users and networks, and among networks. The heterogeneous nature of today's wireless access medium motivates cooperation among different networks, which can benefit both users and service providers. In this brief, we focus on cooperative networking to enhance users QoS through radio resource allocation mechanisms. A literature review is summarized, where the resource allocation mechanisms are classified into single-network and multi-homing ones. The limitations of the existing mechanisms are discussed and a desired cooperative resource allocation framework that can address these limitations is introduced. In the subsequent chapters, cooperative resource aggregation and seamless service provision concepts will be employed to develop an efficient radio resource allocation framework in this heterogeneous networking environment.

Chapter 2
Decentralized Optimal Resource Allocation

Mutli-homing radio resource allocation is considered to be a promising solution that can efficiently exploit the available resources in a heterogeneous wireless access medium to satisfy required QoS, reduce call blocking probability, and enhance mobility support. The main challenge in designing a multi-homing resource allocation algorithm is how to coordinate the allocation from different networks so as to satisfy the user's target QoS while making efficient utilization of available network resources. One simple solution is to employ a central resource manager with a global view over the available resources and the calls required QoS, that can perform the necessary coordination among different networks. However, this solution is not practical in the case that those different networks are operated by different service providers. Hence, the question now is how to coordinate the resource allocation in different networks without a central resource manager. In addition, it is more practical that every network prioritize resource allocation to its own subscribers as compared to other users. In this chapter, we present a decentralized optimal radio resource allocation mechanism that enables each MT to coordinate the resource allocation from different networks to satisfy its target QoS, and allows each network to give a higher priority in allocating its resources to its own subscribers. We first present the system model under consideration, then discuss the problem formulation for the decentralized resource allocation.

2.1 System Model

2.1.1 Wireless Access Networks

Consider a geographical region with a set $\mathcal{N}$ of available wireless access networks, $\mathcal{N} = \{1, 2, \ldots, N\}$. Each network, $n \in \mathcal{N}$, is operated by a unique service provider and has a set, $\mathcal{S}_n$, of BSs/APs in the geographical region with $\mathcal{S}_n = \{1, 2, \ldots, S_n\}$. The BSs/APs of different networks have different coverage that overlaps in some

M. Ismail and W. Zhuang, *Cooperative Networking in a Heterogeneous Wireless Medium*, SpringerBriefs in Computer Science, DOI: 10.1007/978-1-4614-7079-3_2,

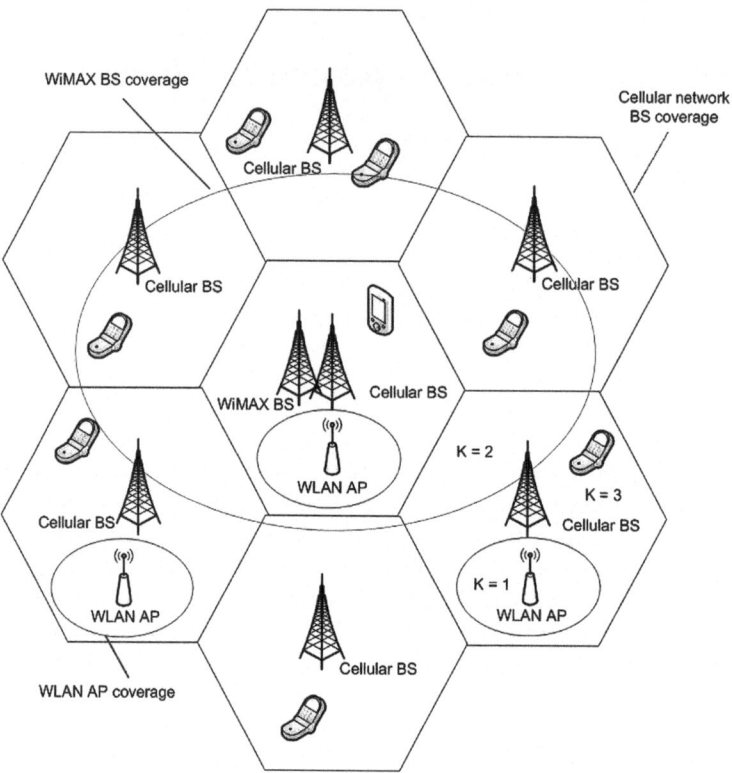

Fig. 2.1 The networks coverage areas

areas. Hence, the geographical region is partitioned to a set $\mathcal{K}$ of service areas, $\mathcal{K} = \{1, 2, \ldots, K\}$. As shown in Fig. 2.1, each service area $k \in \mathcal{K}$ is covered by a unique subset of networks BSs/APs. Each BS/AP, $s \in \mathcal{S}_n$ for $n \in \mathcal{N}$, has a downlink transmission capacity of C_n Mbps.

2.1.2 Network Subscribers and Users

There are M MTs with multiple radio interfaces and multi-homing capabilities in the geographical region, given by the set $\mathcal{M} = \{1, 2, \ldots, M\}$. Each MT has its own home network but can also get service from other available networks. Let $\mathcal{M}_{ns} \subset \mathcal{M}$ denote the set of MTs which lie in the coverage area of the sth BS/AP of the nth network. The set $\mathcal{M}_{ns}$ is further divided into two subsets, $\mathcal{M}_{ns1}$ to denote MTs whose home network is network n, and $\mathcal{M}_{ns2}$ to denote MTs whose home network is not network n. Hence, $\mathcal{M}_{ns1} \cup \mathcal{M}_{ns2} = \mathcal{M}_{ns}$, and $\mathcal{M}_{ns1} \cap \mathcal{M}_{ns2} = \emptyset$. An MT

$m \in \mathcal{M}_{ns1}$ is referred to as network n subscriber, while an MT $m \in \mathcal{M}_{ns2}$ is referred to as network n user.

2.1.3 Service Requests

The MTs service requests are expressed in terms of call required bandwidth. An MT can receive its required bandwidth from all available wireless access networks using its multi-homing capability. The allocated bandwidth from network n to an MT m through BS/AP s in the downlink is given by b_{nms}, with $n \in \mathcal{N}$, $m \in \mathcal{M}_{ns}$, and $s \in \mathcal{S}_n$. Let B be a matrix of bandwidth allocation from each network n through BS/AP s to each MT m, $B = [b_{nms}]$, $n \in \mathcal{N}$, $m \in \mathcal{M}$, $s \in \mathcal{S}_n$, with $b_{nms} = 0$ if MT m is not in the coverage area of network n BS/AP s. It should be noted that, while we study bandwidth allocation in the downlink, the same framework can be applied for bandwidth allocation in the uplink.

The networks support both CBR and VBR services. An MT, m, with a CBR call requires a constant bandwidth B_m from all wireless access networks available at its location. On the other hand, an MT, m, with a VBR call requires a bandwidth allocation within a maximum value $B_m^{\max}$ and a minimum value $B_m^{\min}$. With sufficient available radio resources, the VBR call is allocated its maximum required bandwidth $B_m^{\max}$. When all networks BSs/APs reach their transmission capacity limitation C_n, the allocated bandwidth for the VBR call is degraded towards $B_m^{\min}$ in order to support more calls. Let $\mathcal{M}_{r1}$ denotes the set of MTs in the geographical region with CBR service, while $\mathcal{M}_{r2}$ denotes the set of MTs in the geographical region with VBR service, and both are a subset of $\mathcal{M}$.

We consider call-level radio resource allocation. The radio resource allocation mechanism is to find the optimal resource allocation to a set of MTs in a particular service area from each of the available BSs/APs. As a first step, the resource allocation is performed according to the average call level statistics in different service areas [39]. Hence, a static system is investigated without arrivals of new calls or departures of existing ones. It is assumed that a call admission control procedure is in place [60], and a feasible resource allocation solution exists.

2.2 Formulation of the Radio Resource Allocation Problem

In this section, we discuss the problem formulation of radio resource allocation for a static system of multi-homing MTs in the heterogeneous wireless access medium. A decentralized solution for the problem is then presented based on the problem formulation.

The utility $u_m(b_{nms})$ of network n allocating bandwidth b_{nms} to MT m through BS/AP s is given by

$$u_{nms}(b_{nms}) = \ln(1 + \eta_1 b_{nms}) - \eta_2(1 - p_{nms})b_{nms} \tag{2.1}$$

where η_1 and η_2 are used for scalability of b_{nms} [57], and $p_{nms} \in [0, 1]$ is a priority parameter set by network n BS/AP s on its resources for MT m. The attained network utility from the allocated bandwidth is a concave function of b_{nms} [6] and is given by the first term in the right hand side of (2.1) [39]. The cost that the user pays for the allocated bandwidth is given by the second term in the right hand side of (2.1). This term is a linear function of the allocated bandwidth b_{nms}; hence, the more the allocated bandwidth, the higher the cost. The utility function of (2.1) involves a trade-off between the attained network utility and the cost that the user pays on the network radio resources [28]. The utility function of (2.1) is a concave function of the allocated bandwidth b_{nms} [6]. We employ priority parameter p_{nms} set by network n BS/AP s to MT m to establish service differentiation among different users, which is given by

$$p_{nms} = \begin{cases} 1, & \forall m \in \mathcal{M}_{ns1} \\ \beta, & \forall m \in \mathcal{M}_{ns2} \end{cases} \tag{2.2}$$

where $\beta \in [0, 1)$. Using (2.2) in (2.1), the utility function for a network subscriber accounts only on the attained network utility by that subscriber, while a network user suffers from a trade-off between the attained network utility and the cost that the network sets on its resources [28]. This enables each network to give a higher priority in allocating its resources to its own subscribers than to other users. The allocated bandwidth to MTs with VBR service is reduced, when all networks in the geographical region reach their capacity limitation, in order to support more calls. However, each subscriber should be able to enjoy the resources of its own home network. Hence, it is desirable to differentiate the radio resource allocation performed by a network to its own subscribers and the allocation performed by that network to the other users. This is taken care of by the priority parameter p_{nms} which gives a higher cost on the network resources for the network users than to the network subscribers. Each network, $n \in \mathcal{N}$, assigns a priority parameter value $p_{nms} \in [0, 1)$ on its resources for the users in its coverage area, while setting $p_{nms} = 1$ for its own subscribers. Hence, the subscribers of each network with VBR service enjoy their maximum required bandwidth using their home network radio resources. A network degrades its resource allocation to its own subscribers only so as not to violate the minimum required bandwidth of the other users.

The radio resource allocation objective of each network BS/AP is to maximize the total satisfaction for all MTs that lie within its coverage area, which is given by

$$U_{ns} = \sum_{m \in \mathcal{M}_{ns}} u_{nms}(b_{nms}), \qquad \forall s \in \mathcal{S}_n, n \in \mathcal{N} \tag{2.3}$$

where U_{ns} is the total utility of network n BS/AP s.

The overall radio resource allocation objective of all networks in the geographical region is to find the optimal bandwidth allocation $b_{nms}, \forall n \in \mathcal{N}, \forall m \in \mathcal{M}, \forall s \in \mathcal{S}_n$, which maximizes the total utility in the region, given by

$$U = \sum_{n=1}^{N} \sum_{s=1}^{S_n} U_{ns}. \tag{2.4}$$

The total bandwidth allocation by each network n BS/AP s should be such that the total call traffic load in its coverage area is within the network BS/AP transmission capacity limitation C_n, that is

$$\sum_{m \in \mathcal{M}_{ns}} b_{nms} \leq C_n, \qquad \forall s \in \mathcal{S}_n, n \in \mathcal{N}. \tag{2.5}$$

For an MT with CBR service, the total bandwidth allocation from all available wireless access networks to this MT should satisfy its application required bandwidth, that is

$$\sum_{n=1}^{N} \sum_{s=1}^{S_n} b_{nms} = B_m, \qquad \forall m \in \mathcal{M}_{r1}. \tag{2.6}$$

As for an MT with VBR service, the total bandwidth allocation from all available wireless access networks to this MT should be within the application minimum required bandwidth $B_m^{\min}$ and the application maximum required bandwidth $B_m^{\max}$, that is

$$B_m^{\min} \leq \sum_{n=1}^{N} \sum_{s=1}^{S_n} b_{nms} \leq B_m^{\max}, \qquad \forall m \in \mathcal{M}_{r2}. \tag{2.7}$$

Hence, the radio resource allocation for MTs with multi-homing capabilities in the heterogeneous wireless access medium, for CBR and VBR services, can be expressed by the following optimization problem

$$\begin{aligned} \max_{B \geq 0} \quad & U \\ s.t. \quad & (2.5) - (2.7). \end{aligned} \tag{2.8}$$

Using the utility function definitions in (2.1), (2.3), and (2.4), the objective function of (2.8) is concave and the problem has linear constraints. Therefore, problem (2.8) is a convex optimization problem, and a local maximum is a global maximum as well [6]. Although problem (2.8) can be solved efficiently in polynomial time complexity in a centralized manner using a central resource manager, this is not practical in a case that these networks are operated by different service providers. Thus, it is desirable to develop a decentralized solution of (2.8).

Constraints (2.6) and (2.7) are coupling constraints that make it difficult to obtain the desirable decentralized solution of (2.8) at each network. A decentralized solution can be developed using full dual decomposition of (2.8) [15, 30, 32, 49, 50]. We can rewrite constraint (2.7) in the following form

$$\sum_{n=1}^{N}\sum_{s=1}^{S_n} b_{nms} \le B_m^{\max}, \quad \forall m \in \mathcal{M}_{r2} \tag{2.9}$$

$$\sum_{n=1}^{N}\sum_{s=1}^{S_n} b_{nms} \ge B_m^{\min}, \quad \forall m \in \mathcal{M}_{r2}. \tag{2.10}$$

In order to develop the decentralized solution, first we find the Lagrangian function for (2.8) using constraints (2.9) and (2.10), which can be expressed as

$$L(B, \lambda, \nu, \mu^{(1)}, \mu^{(2)}) = \sum_{n=1}^{N}\sum_{s=1}^{S_n} U_{ns} + \sum_{n=1}^{N}\sum_{s=1}^{S_n} \lambda_{ns} \left(C_n - \sum_{m \in \mathcal{M}_{ns}} b_{nms} \right)$$

$$+ \sum_{m \in \mathcal{M}_{r1}} \nu_m \left(B_m - \sum_{n=1}^{N}\sum_{s=1}^{S_n} b_{nms} \right)$$

$$+ \sum_{m \in \mathcal{M}_{r2}} \mu_m^{(1)} \left(B_m^{\max} - \sum_{n=1}^{N}\sum_{s=1}^{S_n} b_{nms} \right)$$

$$+ \sum_{m \in \mathcal{M}_{r2}} \mu_m^{(2)} \left(\sum_{n=1}^{N}\sum_{s=1}^{S_n} b_{nms} - B_m^{\min} \right) \tag{2.11}$$

with $\lambda = (\lambda_{ns} : n \in \mathcal{N}, s \in \mathcal{S}_n)$ defined to be a matrix of Lagrange multipliers corresponding to capacity constraint (2.5), and $\lambda_{ns} \ge 0$, $\nu = (\nu_m : m \in \mathcal{M}_{r1})$, $\mu^{(1)} = (\mu_m^{(1)} : m \in \mathcal{M}_{r2})$, $\mu^{(2)} = (\mu_m^{(2)} : m \in \mathcal{M}_{r2})$ are vectors of lagrange multipliers corresponding to the required bandwidth constraints (2.6), (2.9), and (2.10) respectively, and $\mu_m^{(1)}, \mu_m^{(2)} \ge 0$. The dual function is given by

$$h(\lambda, \nu, \mu^{(1)}, \mu^{(2)}) = \max_{B \ge 0} L(B, \lambda, \nu, \mu^{(1)}, \mu^{(2)}) \tag{2.12}$$

and the dual problem corresponding to the primal problem (2.8) is expressed by

$$\min_{(\lambda, \mu^{(1)}, \mu^{(2)}) \ge 0, \nu} h(\lambda, \nu, \mu^{(1)}, \mu^{(2)}). \tag{2.13}$$

A strong duality holds since the optimization problem (2.8) is a convex optimization problem, which makes the optimal values for the primal and dual problems equal [6]. The maximization problem (2.12) can be written as

$$h(\lambda, v, \mu^{(1)}, \mu^{(2)}) = \sum_{n=1}^{N} \sum_{s=1}^{S_n} \max_{B \geq 0} \left\{ U_{ns} - \lambda_{ns} \sum_{m \in \mathcal{M}_{ns}} b_{nms} \right.$$

$$\left. - \sum_{m \in \mathcal{M}_{r1}} v_m b_{nms} - \sum_{m \in \mathcal{M}_{r2}} (\mu_m^{(1)} - \mu_m^{(2)}) b_{nms} \right\}. \quad (2.14)$$

Then, each network BS/AP can solve its own network utility maximization (NUM) problem, given by

$$\max_{B \geq 0} \left\{ U_{ns} - \lambda_{ns} \sum_{m \in \mathcal{M}_{ns}} b_{nms} - \sum_{m \in \mathcal{M}_{r1}} v_m b_{nms} - \sum_{m \in \mathcal{M}_{r2}} (\mu_m^{(1)} - \mu_m^{(2)}) b_{nms} \right\}. \quad (2.15)$$

By applying the Karush-Kuhn-Tucker (KKT) conditions on (2.15), each network BS/AP can find the bandwidth allocation, b_{nms}, for fixed values of λ, v, $\mu^{(1)}$, and $\mu^{(2)}$. Thus, we have

$$\frac{\partial U_{ns}}{\partial b_{nms}} - \lambda_{ns} - v_m - (\mu_m^{(1)} - \mu_m^{(2)}) = 0 \quad (2.16)$$

which results in

$$b_{nms} = \left[\left(\frac{\eta_1}{\lambda_{ns} + v_m + \eta_2(1 - p_{nms})} - 1 \right) / \eta_1 \right]^{+}, \quad \forall m \in \mathcal{M}_{r1} \quad (2.17)$$

$$b_{nms} = \left[\left(\frac{\eta_1}{\lambda_{ns} + (\mu_m^{(1)} - \mu_m^{(2)}) + \eta_2(1 - p_{nms})} - 1 \right) / \eta_1 \right]^{+}, \quad \forall m \in \mathcal{M}_{r2} \quad (2.18)$$

where the notion $[\cdot]^{+}$ is a projection on the positive quadrature to account for the fact that $B \geq 0$. By solving the dual problem (2.13), we can obtain the optimal values for the Lagrange multipliers that results in the optimal bandwidth allocation b_{nms} of (2.17) and (2.18). For a fixed bandwidth allocation B, the dual problem can be written as

$$\sum_{n=1}^{N} \sum_{s=1}^{S_n} \min_{\lambda \geq 0} \left\{ \lambda_{ns} \left(C_n - \sum_{m \in \mathcal{M}_{ns}} b_{nms} \right) \right\} + \sum_{m \in \mathcal{M}_{r1}} \min_{v} \left\{ v_m \left(B_m - \sum_{n=1}^{N} \sum_{s=1}^{S_n} b_{nms} \right) \right\}$$

$$+ \sum_{m \in \mathcal{M}_{r2}} \min_{\mu^{(1)} \geq 0} \left\{ \mu_m^{(1)} \left(B_m^{\max} - \sum_{n=1}^{N} \sum_{s=1}^{S_n} b_{nms} \right) \right\}$$

$$+ \sum_{m \in \mathcal{M}_{r2}} \min_{\mu^{(2)} \geq 0} \left\{ \mu_m^{(2)} \left(\sum_{n=1}^{N} \sum_{s=1}^{S_n} b_{nms} - B_m^{\min} \right) \right\}. \quad (2.19)$$

For a differentiable dual function, a gradient descent method can be applied so as to find the optimal values for the Lagrangian multipliers [6], which is given by

$$\lambda_{ns}(i+1) = \left[\lambda_{ns}(i) - \alpha_1 \left(C_n - \sum_{m \in \mathcal{M}_{ns}} b_{nms}(i)\right)\right]^+ \tag{2.20}$$

$$\nu_m(i+1) = \nu_m(i) - \alpha_2 \left(B_m - \sum_{n=1}^{N}\sum_{s=1}^{S_n} b_{nms}(i)\right) \tag{2.21}$$

$$\mu_m^{(1)}(i+1) = \left[\mu_m^{(1)}(i) - \alpha_3 \left(B_m^{\max} - \sum_{n=1}^{N}\sum_{s=1}^{S_n} b_{nms}(i)\right)\right]^+ \tag{2.22}$$

$$\mu_m^{(2)}(i+1) = \left[\mu_m^{(2)}(i) - \alpha_4 \left(\sum_{n=1}^{N}\sum_{s=1}^{S_n} b_{nms}(i) - B_m^{\min}\right)\right]^+ \tag{2.23}$$

where i is an iteration index and α_j, $j = \{1, 2, 3, 4\}$, is a fixed sufficiently small step size. As the gradient of (2.19) satisfies the Lipchitz continuity condition, the convergence of (2.20)–(2.23) towards the optimal solution is guaranteed [6]. Hence, the radio resource allocation b_{nms} of (2.17) and (2.18) converges to the optimal solution.

2.3 A Decentralized Optimal Resource Allocation (DORA) Algorithm

The decomposition approach for optimization problem (2.8) is defined in two levels. The first one is a lower level that is executed at each network, $n \in \mathcal{N}$, BS/AP, $s \in \mathcal{S}_n$, so as to find the optimal radio resource allocation b_{nms} for each MT $m \in \mathcal{M}_{ns}$. This optimal resource allocation is found by solving the sub-problems given in (2.15) by BSs/APs, which results in the solution of (2.17) for MTs with CBR service and (2.18) for MTs with VBR service. The other is a higher level, where the master problem is solved. The master problem is given in (2.19) and its optimal solution is obtained using the iterative method introduced in (2.20)–(2.23). The role of the master problem is to set the dual variables λ, ν, $\mu^{(1)}$ and $\mu^{(2)}$ so as to coordinate the sub-problems solution at each network BS/AP. This is illustrated in Fig. 2.2.

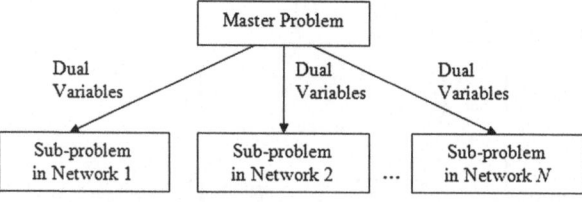

Fig. 2.2 Decomposition of problem (2.8)

Following the classical interpretation of λ_{ns} in economics as the resource price [32], we refer to λ_{ns} as the link access price for network n BS/AP s. Basically, λ_{ns} serves as an indication of the capacity limitation experienced by network n link resources in BS/AP s. Hence, when the total call traffic load in network n BS/AP s ($\sum_{m \in \mathcal{M}_{ns}} b_{nms}$) reaches the capacity limitation (C_n), the link access price (λ_{ns}) increases to denote that it is expensive to use that link. The rest of the Lagrangian multipliers, namely ν_m which is used by MTs with CBR service, and $\mu_m^{(1)}$ and $\mu_m^{(2)}$ which are used by MTs with VBR service, are coordination parameters. Hence, ν_m is used by MT m to coordinate the allocations by the available BSs/APs so as to ensure that the required bandwidth B_m is met. Similarly, $\mu_m^{(1)}$ and $\mu_m^{(2)}$ are used by MT m to coordinate the BS/AP resource allocations of different networks so as to ensure that the allocated resources lie within the specified required bandwidth range $[B_m^{\min}, B_m^{\max}]$.

The link access price λ_{ns} is calculated at each network BS/AP according to its capacity limitation and the total call traffic load experienced by the BS/AP. The coordination parameter ν_m is calculated at each MT with CBR service, while the coordination parameters $\mu_m^{(1)}$ and $\mu_m^{(2)}$ are calculated by each MT with VBR service. All coordination parameters are calculated based on the allocated bandwidth from different wireless access networks and the MT required bandwidth. The decentralized optimal radio resource allocation (DORA) algorithm can be explained using the scenario given in Fig. 2.3. Consider an MT which lies in the coverage area of a WLAN AP and cellular network and WiMAX BSs. Each BS/AP defines an initial feasible value for its link access price λ_{ns}. Similarly, the MT defines an initial feasible value for its coordination parameter(s). Each BS/AP performs its bandwidth allocation to the MT based on the network BS/AP link access price, the MT priority parameter and its coordination parameter values. Each BS/AP then updates its link access price value based on its capacity limitation and its experienced total call traffic load (due to the previous iteration resource allocation). Also, the MT updates its coordination parameter(s) (ν_m for MT with CBR service and $\mu_m^{(1)}$ and $\mu_m^{(2)}$ for MT with VBR service) based on the difference between its required bandwidth and the previous iteration total resource allocation. The updated coordination parameter for the new iteration (ν_m or the difference $\mu_m^{(1)} - \mu_m^{(2)}$) is broadcasted by the MT to the different available wireless access networks through the MT different radio interfaces so as to coordinate the resource allocation from different networks. As a result, each BS/AP can update its bandwidth allocation to the MT (using the updated link access price and coordination parameter values). The process continues over a number of iterations until the MT required bandwidth can be met eventually.

The detailed (DORA) algorithm is given in Table 2.1, where ψ is a small tolerance.

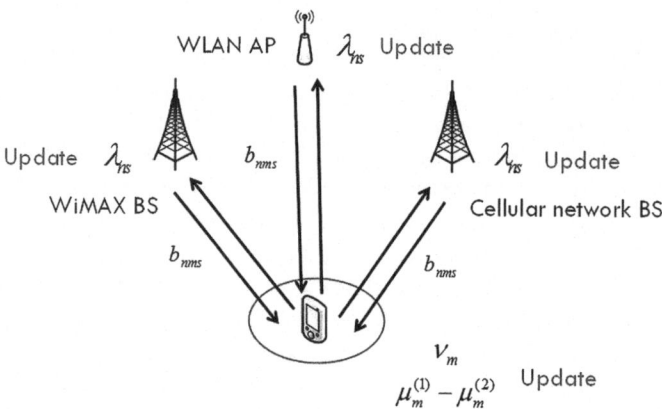

Fig. 2.3 Decentralized radio resource allocation

2.4 Numerical Results and Discussion

This section presents numerical results for the radio resource allocation problem (2.8) using the DORA algorithm given in Table 2.1. We consider a simplified system model with a geographical region that is entirely covered by an IEEE 802.16e WiMAX BS and partially covered by a 3G cellular network BS and an IEEE 802.11b WLAN AP [39], as shown in Fig. 2.4. Thus, $\mathcal{N} = \{1, 2, 3\}$, with the WiMAX, cellular network, and WLAN indexed as 1, 2, and 3 respectively. Each network has only one BS/AP in the geographical region, i.e. $S_n = \{1\}$, $\forall n \in \mathcal{N}$. As a result, the geographical region is described by three service areas, $\mathcal{K} = \{1, 2, 3\}$. In service area 1, only the WiMAX BS coverage is available. In service area 2, both the WiMAX and cellular network coverages are available. In service area 3, all three networks are available. The transmission capacities of the three networks are given by $C_1 = 20$ Mbps, $C_2 = 2$ Mbps, and $C_3 = 11$ Mbps.

For the priority mechanism, different networks can set different costs on their resources through the priority parameter p_{nms}. As the cellular network has the lowest transmission capacity among all the available networks, it sets the highest cost on its resources so as to devote them to its own subscribers. Both the WiMAX and WLAN have a high transmission capacity, however, the WiMAX covers a larger area with more users. Hence, the WiMAX sets a higher cost on its resources than the WLAN with its limited coverage area. So, for network users we set $p_{1m1} = 0.6$, $p_{2m1} = 0.5$, and $p_{3m1} = 0.8$.

Let the required bandwidth allocation be 256 Kbps for an MT with CBR service, while for an MT with VBR service the required bandwidth allocation lies in the range $[256, 512]$ Kbps. Let the number of subscribers for network n in service area k with service r be M_{nkr} with $r = 1$ for CBR service and $r = 2$ for VBR service. We vary the number of WLAN subscribers with CBR calls in service area 3 (M_{331}) and fix all other parameters to study the system performance. The number of different network subscribers in all service areas are given in Table 2.2.

Figures 2.5, 2.6, 2.7, and 2.8 depict various bandwidth allocation results versus the number of ongoing CBR calls for the WLAN subscribers in service area 3 (M_{331}).

Figure 2.5 shows the total allocated bandwidth by each network BS/AP. Both the WiMAX and cellular network BSs reach their capacity limitation, independent of

Table 2.1 DORA Algorithm

1: **Input:** $C_{ns} \ \forall n \in \mathcal{N}, \ \forall s \in \mathcal{S}_n, \ B_m \ \forall m \in \mathcal{M}_{r1}, \ [B_m^{\min}, B_m^{\max}] \ \forall m \in \mathcal{M}_{r2};$
2: **Initialization:** $i \longleftarrow 1; \ \lambda_{ns}(1) \geq 0; \ v_m(1); \ \mu_m^{(1)}(1) \geq 0; \ \mu_m^{(2)}(1) \geq 0, \ b_{nms}(0) = \{\}, \ j = 0;$
3: **while** $j = 0$ **do**
4:     **for** $n \in \mathcal{N}$ **do**   // Bandwith Allocation at Each Network BS/AP
5:         **for** $m \in \mathcal{M}$ **do**
6:             **for** $s \in \mathcal{S}_n$ **do**
7:                 **if** $m \in \mathcal{M}_{ns}$ **then**
8:                     $b_{nms}(i) = [(\frac{\eta_1}{\lambda_{ns}(i)+v_m(i)+\eta_2(1-p_{nms})} - 1)/\eta_1]^+, \qquad m \in \mathcal{M}_{r1};$
9:                     $b_{nms}(i) = [(\frac{\eta_1}{\lambda_{ns}(i)+(\mu_m^{(1)}(i)-\mu_m^{(2)}(i))+\eta_2(1-p_{nms})} - 1)/\eta_1]^+,$
10:                                 $m \in \mathcal{M}_{r2};$
11:                 **end if**
12:             **end for**
13:         **end for**
14:     **end for**
15:     **if** $|b_{nms} - 1| > \psi$ **then**
16:         **for** $n \in \mathcal{N}$ **do**   // Update of Link Access Price at Each Network BS/AP
17:             **for** $s \in \mathcal{S}_n$ **do**
18:                 $\lambda_{ns}(i+1) = [\lambda_{ns}(i) - \alpha_1(C_{ns} - \sum_{m \in \mathcal{M}_{ns}} b_{nms}(i))]^+;$
19:             **end for**
20:         **end for**
21:         **for** $m \in \mathcal{M}$ **do**   // Update of Coordination Parameters at Each MT
22:             $v_m(i+1) = v_m(i) - \alpha_2(B_m - \sum_{n=1}^{N} \sum_{s=1}^{S_n} b_{nms}(i)), \qquad \forall m \in \mathcal{M}_{r1}$
23:             $\mu_m^{(1)}(i+1) = [\mu_m^{(1)}(i) - \alpha_3(B_m^{\max} - \sum_{n=1}^{N} \sum_{s=1}^{S_n} b_{nms}(i))]^+, \qquad \forall m \in \mathcal{M}_{r2};$
24:             $\mu_m^{(2)}(i+1) = [\mu_m^{(2)}(i) - \alpha_4(\sum_{n=1}^{N} \sum_{s=1}^{S_n} b_{nms}(i) - B_m^{\min})]^+, \qquad \forall m \in \mathcal{M}_{r2};$
25:         **end for**
26:         $i \longleftarrow i+1$
27:     **else**
28:         $j = 1;$
29:     **end if**
30: **end while**
31: **Output:** B^*.

M_{331}. On the other hand, the WLAN AP increases its total allocated bandwidth with M_{331} so as to accommodate more subscribers. The WLAN AP reaches its capacity limitation at $M_{331} = 14$.

In the following results, we study the total allocated bandwidth from each network BS/AP to subscribers of different networks in all three service areas.

Figure 2.6a shows the total allocated bandwidth by each network BS/AP for the CBR WLAN subscribers in service area 3. Because of the priority mechanism, the WLAN AP supports its own subscribers with all their required bandwidth in order to avoid the associated high cost of the BSs resources of WiMAX and cellular network. Hence, The bandwidth allocation for the WLAN subscribers from the WiMAX (M-L) and cellular network (C-L) BSs is equal to zero, while the WLAN AP allocated bandwidth (L-L) increases with M_{331} so as to accommodate more subscribers. For $M_{331} > 34$, there is no sufficient resources at the WLAN AP to support individ-

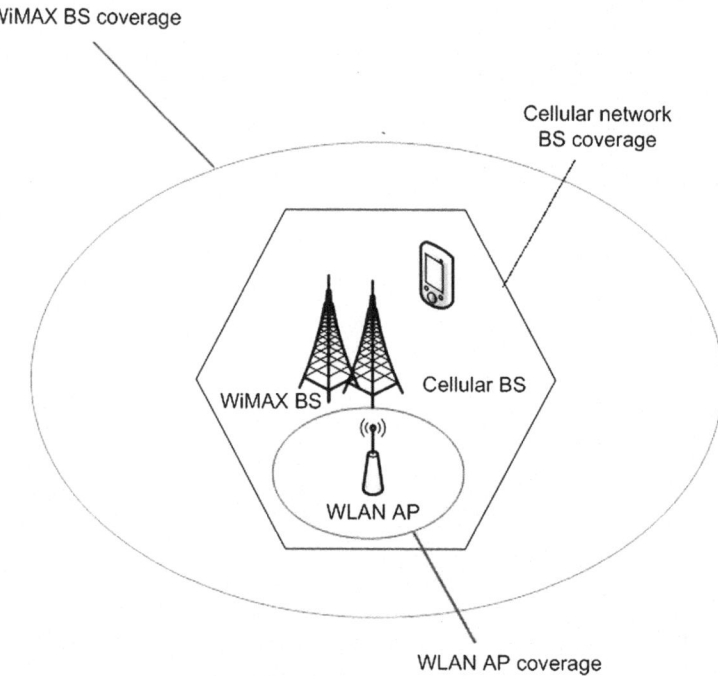

Fig. 2.4 Service areas under consideration

Table 2.2 Number of subscribers of different networks in different service areas

Parameter	Value	Parameter	Value	Parameter	Value	Parameter	Value
M_{111}	10	M_{122}	7	M_{221}	8	M_{232}	5
M_{112}	10	M_{131}	5	M_{222}	8	M_{332}	5
M_{121}	7	M_{132}	5	M_{231}	5	M_{331}	Variable

ually its own subscribers. Hence, the WiMAX BS increases its bandwidth allocation to support the WLAN subscribers. The support comes only from the WiMAX BS as it sets a lower cost on its resources than the cellular network BS.

Figure 2.6b shows the allocated bandwidth by each network BS/AP for the VBR WLAN subscribers in service area 3. For $M_{331} \geq 22$, the WLAN AP decreases its allocated bandwidth to the VBR subscribers (L-L) in order to support the increasing number of the CBR subscribers. This is compensated by an increase in the bandwidth allocation from the WiMAX BS (M-L) in order to keep the total bandwidth allocation constant at the call maximum required bandwidth (512 Kbps for each VBR call). For $M_{331} > 27$, any further increase in the bandwidth allocation from the WiMAX BS to the WLAN subscribers would degrade the WiMAX BS bandwidth allocation to its own VBR subscribers. This is not allowed, however, by the priority mechanism as it gives higher priority on the WiMAX BS resources to the WiMAX subscribers. Hence, the WiMAX BS decreases its allocated bandwidth to the VBR WLAN subscribers which reduces the VBR call total bandwidth allocation towards the call minimum required bandwidth. For $M_{331} > 34$, the WLAN AP decreases its bandwidth allocation to its VBR subscribers in order to support the increasing number of its CBR subscribers. Hence, the WiMAX BS increases its bandwidth allocation to the WLAN VBR subscribers so as not to violate their minimum required bandwidth (256 Kbps for each VBR call).

Figure 2.7a shows the total allocated bandwidth by each network BS/AP to the cellular network subscribers, with CBR and VBR calls, in service area 3. The total allocated bandwidth of CBR cellular network subscribers (C-CBR Total) comes from

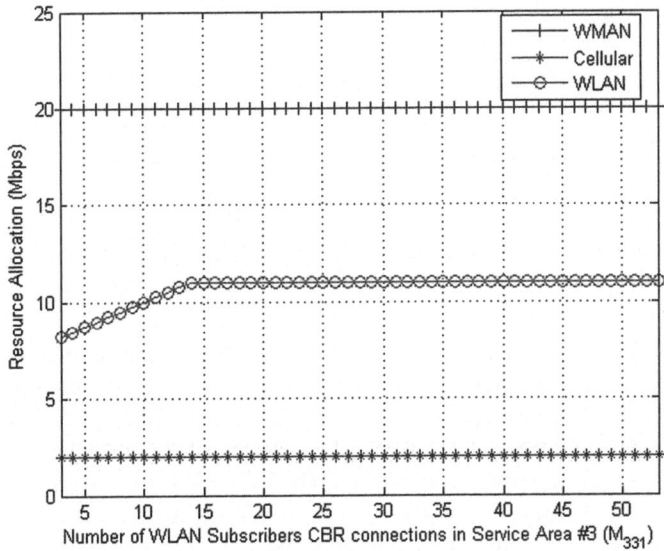

Fig. 2.5 Total bandwidth allocation by each network BS/AP

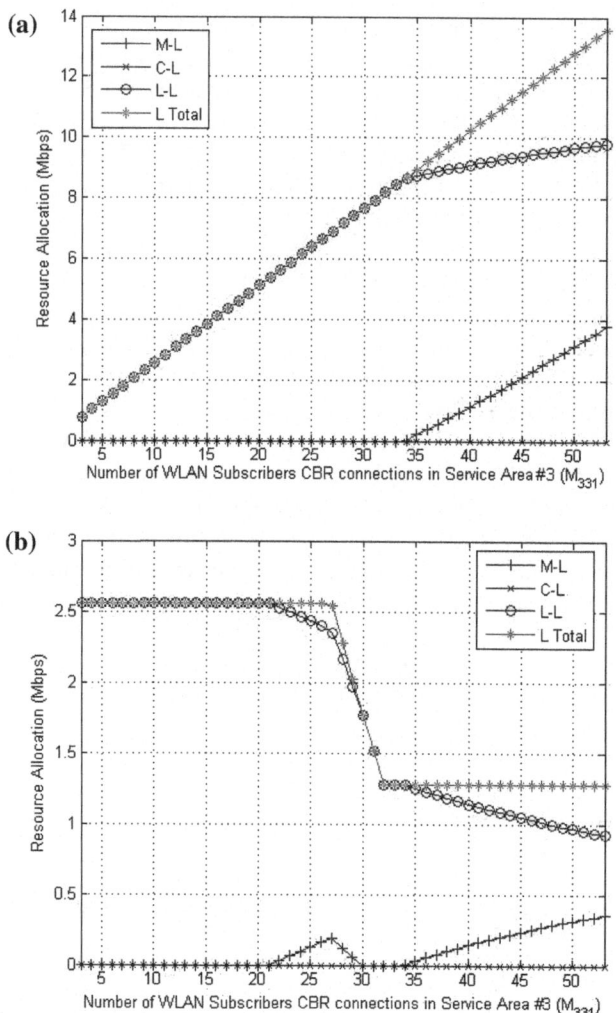

Fig. 2.6 Total bandwidth allocation by each network BS/AP to **a** CBR and **b** VBR WLAN subscribers

the WLAN AP (L-C-CBR). The allocated bandwidth from the cellular network BS (C-C-CBR) is zero, as it uses its bandwidth to support its own subscribers in service area 2 (which is covered only by the cellular network BS, and the WiMAX BS with a higher cost for bandwidth). As for the WiMAX BS zero bandwidth allocation (M-C-CBR), it is due to the higher cost that the WiMAX BS sets on its resources as compared to the WLAN AP. For $M_{331} > 18$, the WLAN AP decreases its bandwidth allocation to the CBR cellular network subscribers in order to support its increasing number of subscribers (M_{331}). Hence, the WiMAX BS increases its allocation to the CBR cellular network subscribers in order to keep the total bandwidth allocation

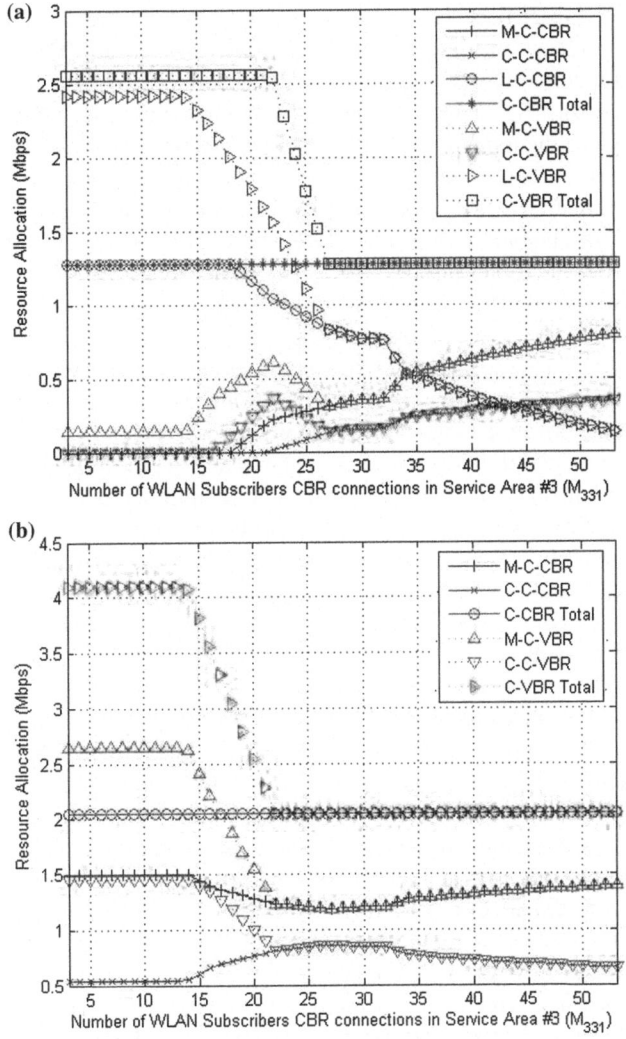

Fig. 2.7 Total bandwidth allocation by each network BS/AP to the cellular network subscribers in **a** Area 3 and **b** Area 2

constant at the required bandwidth (256 Kbps for each CBR call). For $M_{331} > 21$, more allocated bandwidth is required from the WiMAX BS to keep the CBR cellular network subscriber total allocation constant; however, this would increase the associated cost due to the WiMAX BS low priority parameter for the network users. Hence, the cellular network BS increases its allocated bandwidth to support its own CBR subscribers. As shown in the figure, the total bandwidth allocation is always constant at the call required bandwidth. For the VBR subscribers, the WLAN AP decreases its bandwidth allocation to the VBR cellular network subscribers with M_{331} in order to support its own subscribers. This is compensated for by an increase in the WiMAX BS bandwidth allocation to keep the total allocated bandwidth (C-VBR-Total) at its maximum required bandwidth (512 Kbps for each VBR call). For $M_{331} > 17$, the cellular network BS increases its bandwidth allocation to its VBR subscribers in order to reduce the amount of required bandwidth from the WiMAX BS due to the associated high cost. For $M_{331} > 22$, any further increase in the allocated bandwidth from the WiMAX BS to the VBR cellular network subscribers would reduce the WiMAX BS allocation to its own VBR subscribers. Hence, the WiMAX BS decreases its allocated bandwidth to the VBR cellular network subscribers. Also, the cellular network BS decreases its allocated bandwidth to its VBR subscribers to support its CBR subscribers in this area. As a result, the total allocated bandwidth to the VBR cellular network subscribers starts to decrease towards the minimum required bandwidth. For $M_{331} > 26$, the WiMAX and cellular network BSs increase their bandwidth allocation to the VBR cellular network subscribers in order to compensate for the reduction in the allocated bandwidth from the WLAN AP and keep the total bandwidth allocation constant at the call minimum required bandwidth.

Figure 2.7b shows the total allocated bandwidth by each network BS/AP to the cellular network subscribers in service area 2. The allocated bandwidth comes only from the WiMAX and cellular network BSs since the MTs are out of the coverage area of the WLAN AP. For the CBR subscribers with $M_{331} > 14$, the WiMAX BS reduces its allocated bandwidth to the CBR cellular network subscribers to support its own subscribers with their maximum required bandwidth. As a result, the cellular network BS increases its allocated bandwidth. For $M_{331} > 32$, the cellular network BS reduces its bandwidth allocation to support its subscribers in area 3 (refer to Fig. 2.7a). This is compensated for by an increase in the WiMAX BS allocated bandwidth to the CBR cellular network subscribers. In all the cases, the total bandwidth allocation (C-CBR Total) is constant at the required bandwidth (256 kbps for each CBR user). For the VBR subscribers with $M_{331} > 14$, the cellular network BS cannot further keep its VBR subscribers in area 2 at their maximum required bandwidth, and has to decrease its allocated bandwidth to support the CBR cellular network subscribers in this area. Also, the WiMAX BS has to decrease its bandwidth allocation to satisfy its own VBR subscribers with their maximum required bandwidth. Therefore, the total bandwidth allocation (C-VBR Total) starts to decrease towards the minimum required bandwidth. As in the CBR bandwidth allocation, for $M_{331} > 32$, the cellular network BS reduces its allocated bandwidth to its VBR subscribers in area 2 to support its subscribers in area 3. As a result, the WiMAX BS increases its bandwidth allocation to keep the total allocated bandwidth constant at the minimum required bandwidth.

Figure 2.8a shows the total allocated bandwidth by each network BS/AP to the WiMAX subscribers in service area 3. For both CBR and VBR calls, most of the allocated bandwidth comes from the WiMAX BS (M-M-CBR and M-M-VBR), so as to reduce the associated cost of the WLAN bandwidth allocation. The allocated bandwidth from the cellular network BS (C-M-CBR and C-M-VBR) is zero, as it allocates radio resources to its own subscribers in service areas 2 and 3. For $M_{331} > 13$, the WLAN AP decreases its allocated bandwidth to the VBR WiMAX subscribers in order to support its own subscribers. Hence, the WiMAX BS increases its allocated

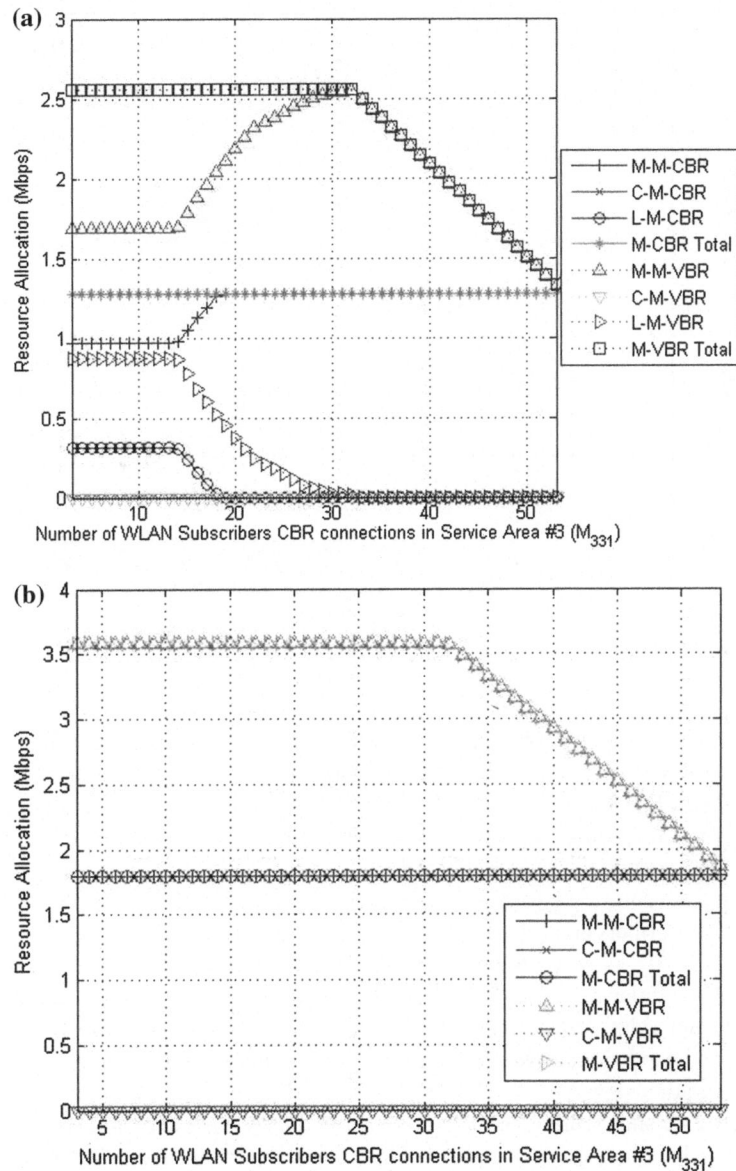

Fig. 2.8 Total bandwidth allocation by each network BS/AP to the WiMAX subscribers in **a** Area 3, **b** Area 2, and **c** Area 1

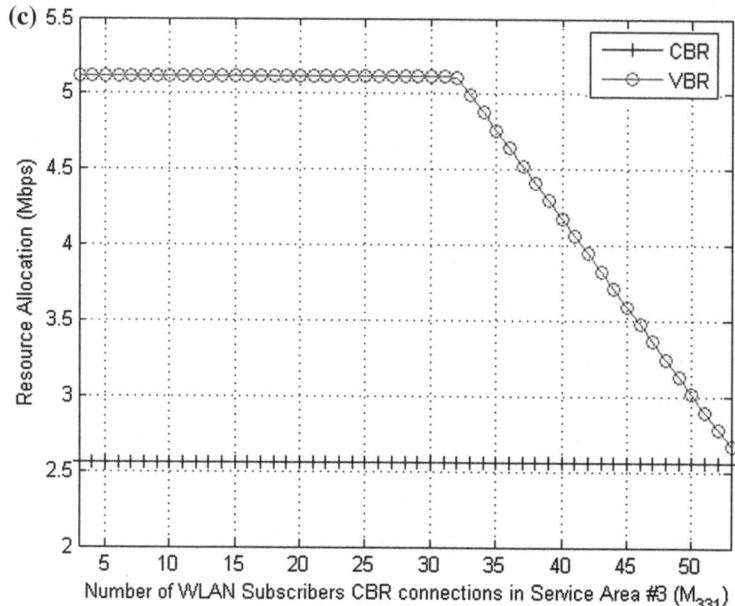

Fig. 2.8 (Continued)

bandwidth to support its own subscribers. For $M_{331} > 18$, all the required bandwidth to support the CBR calls (M-CBR-Total) in service area 3 comes from the
WiMAX BS. For $M_{331} > 32$, the WiMAX BS reduces its bandwidth allocation to
the VBR WiMAX subscribers towards the minimum required bandwidth to support
the WLAN subscribers (refer to Fig. 2.6).

Figure 2.8b shows the total allocated bandwidth by each network BS/AP to the
WiMAX subscribers in service area 2. The total allocated bandwidth comes only from
the WiMAX BS (M-M-CBR and M-M-VBR) although the MTs lie in the coverage
area of the cellular network. This is due to the associated high cost of the cellular
network bandwidth. Again, as in Fig. 2.8a, for $M_{331} > 32$, the WiMAX BS decreases
its allocated bandwidth to the VBR subscribers to support the WLAN subscribers in
service area 3.

Figure 2.8c shows the total allocated bandwidth by each network BS/AP to the
WiMAX subscribers in service area 1. Since the MTs are outside the coverage areas
of the cellular network BS and WLAN AP, the total bandwidth allocation comes only
from the WiMAX BS. For $M_{331} > 32$, the WiMAX BS allocated bandwidth to the
VBR calls is reduced to support the WLAN subscribers in area 3.

From the results in Figs. 2.6, 2.7, and 2.8, service degradation of VBR calls starts
from the cellular network subscribers as these users depend heavily on other networks
in order to satisfy their required bandwidth. Because of the priority mechanism, these
networks give higher priority in allocating their resources to their own subscribers,

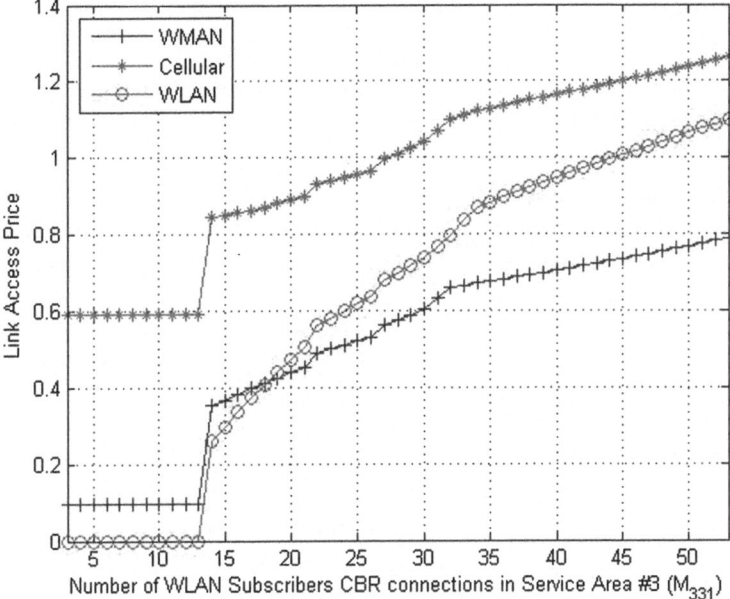

Fig. 2.9 Link access price

leading to a reduced bandwidth allocated to the VBR calls of cellular network subscribers.

Figure 2.9 shows the variation in the link access price (λ_{ns}). For $M_{331} < 14$, the WLAN AP has not yet reached its capacity limitation, resulting in its link access price value equal to zero. On the other hand, the WiMAX and the cellular network BSs have a high value of link access price as they reach their capacity limitation (refer to Fig. 2.5). The cellular network BS has the highest link access price value due to its lowest capacity. For $M_{331} \geq 14$, the BSs/AP of three networks reach their capacity limitation. This calls for a higher link access prices for all three networks. As M_{331} increases, the link access price value increases to indicate that it is more expensive to use these links. These results follow the complementary slackness condition [6]. Normally, the WLAN AP has a lower link access price than the WiMAX BS, since the number of users supported by the WLAN AP is less than those supported by the WiMAX BS in the three areas. But as the WLAN AP gives a lower cost on its resources using the priority parameter p_{3m1}, most of the users in area 3 use its bandwidth, and the WLAN subscribers in area 3 are mainly supported by the WLAN AP, which causes the link access price for the WLAN AP to increase above the link access price value of the WiMAX BS for $M_{331} > 18$.

2.5 Summary

In this chapter, a decentralized optimal resource allocation (DORA) algorithm in a heterogeneous wireless access environment is presented. The algorithm has the following features:

1. It is a decentralized algorithm. Each network BS/AP solves its own NUM problem and performs its resource allocation. No central resource manager is required.
2. It supports MTs with multi-homing capabilities for multi-services, namely, CBR and VBR services.
3. It allows for service differentiation, among the network subscribers and the other users. As a result, the network subscribers enjoy their maximum required bandwidth using their home network resources.
4. The MTs play an active role in the resource allocation operation by coordinating the available wireless access networks to satisfy their required bandwidth.

The algorithm is limited to a static system with no arrival of new calls or departure of existing ones with the objective of identifying the role of different network entities in such a decentralized architecture model. In the next chapter, we discuss the main limitations of the DORA algorithm in a dynamic system with call arrivals and departures and present some modifications to address these limitations.

Chapter 3
Prediction Based Resource Allocation

In a dynamic environment, call arrivals and departures in different service areas may trigger reallocations for all MTs in service. In a decentralized architecture, this is translated to a heavy signalling overhead between the MTs and different BSs/APs with every call arrival and/or departure in any service area. Hence, the main challenge is how to develop an efficient decentralized radio resource allocation mechanism that reduces the associated signalling overhead with call arrivals and departures. In this chapter, concepts of call traffic load prediction and network cooperation are introduced to address the challenges that face the decentralized resource allocation in a dynamic environment.

3.1 Introduction

In Chap. 2, the DORA algorithm is presented to support MTs with multi-homing capabilities in a heterogeneous wireless access medium. The DORA algorithm mainly identifies the role of different entities in the heterogeneous wireless access medium in order to enable a decentralized architecture. Specifically, the main role of a network, $n \in \mathcal{N}$, BS/AP, $s \in \mathcal{S}_n$, in the decentralized architecture is to update a link access price value (λ_{ns}) that indicates the capacity limitation experienced by this BS/AP. On the other hand, the main role of an MT, m, is to update its coordination parameter(s) (ν_m for MT with CBR service, or $\mu_m^{(1)} - \mu_m^{(2)}$ for MT with VBR service) in order to satisfy its required bandwidth. Both link access price values for different BSs/APs and coordination parameter(s), together with the priority parameter p_{nms}, determine the allocated resources from each network BS/AP so as to satisfy the MT total required bandwidth. The DORA algorithm is an iterative one that relies on signalling exchange between an MT and different BSs/APs in order to reach the optimal resource allocation from each BS/AP to the MT. This includes the exchange of the current iteration MT coordination parameter (from MT to BSs/APs) and the corresponding BS/AP resource allocation b_{nms} (from each BS/AP to MT).

M. Ismail and W. Zhuang, *Cooperative Networking in a Heterogeneous Wireless Medium*, SpringerBriefs in Computer Science, DOI: 10.1007/978-1-4614-7079-3_3, © The Author(s) 2013

The DORA algorithm is proposed for a static environment without arrival of new calls or departure of existing ones.

As illustrated in Fig. 2.9, due to the complimentary slackness condition for (2.8), we have the following observations:

1. When the total call traffic load ($\sum_{m \in \mathcal{M}_{ns}} b_{nms}$) carried by network n BS/AP s is less than the BS/AP transmission capacity limitation C_n, the corresponding optimal link access price value $\lambda_{ns}^* = 0$. This results in allocating the maximum required bandwidth for all VBR calls under this BS/AP jurisdiction.
2. When the carried call traffic load reaches the BS/AP transmission capacity limitation, $\lambda_{ns}^* > 0$. Hence, the allocated bandwidth to each of the VBR calls in service is reduced towards the call minimum required bandwidth so as to support new incoming calls.

In a dynamic system, with call arrivals to and departures from different service areas, the carried call traffic load by each BS/AP fluctuates over time. This in turn results in a fluctuating (time-varying) optimal value for the link access price λ_{ns}^* and hence a fluctuating bandwidth allocation matrix B^*, with every call arrival or departure. This results in the following limitations for the DORA algorithm [24]:

1. A fluctuating link access price λ_{ns} triggers bandwidth reallocations to the existing calls. Let I denote the number of iterations required by the DORA algorithm to reach the optimal resource allocation. In the decentralized architecture, information exchange between MTs and BSs/APs for coordination parameter updates is required for the I iterations in order to support an optimal bandwidth reallocation. This signalling exchange should take place, for all MTs in service, with every call arrival to or departure from any service area k. In general, the signalling overhead is a function of the call arrival and departure rates, the numbers of existing calls in different service areas, and the number of iterations required for the algorithm to converge to the optimal resource allocation. Hence, excessive signalling overhead is needed for information exchange between existing MTs and different BSs/APs, which makes the DORA algorithm too expensive to implement in a dynamic system.
2. Due to the random nature of call arrivals and departures in different service areas, it is possible that an arrival or departure event occurs during the I iterations. Hence, the DORA algorithm may not converge in practice to an optimal resource allocation.
3. The signalling information exchange for the I iterations between an MT and different BSs/APs takes place on both up and down links. Let σ denote the total time duration of the signalling exchange for the I iterations. When there exists a network with contention based medium access control among the available wireless networks, it is expected that σ increases with the call arrival rates as more MTs will be involved in the signalling procedure. Hence, the DORA algorithm can lead to high handoff latency, which is not desirable for seamless service provision.

In this chapter, we aim to extend the DORA algorithm so as to account for the system dynamics in terms of call arrivals and departures, and hence to perform an efficient radio resource allocation. We set two objectives for this resource allocation: (1) to significantly reduce the required resource reallocations to existing calls and the associated signalling overhead over the air interface in the decentralized network architecture, with call arrivals to and departures from different service areas; and (2) to achieve an acceptable call blocking probability and a sufficient amount of allocated resources per VBR call. These objectives are achieved through a prediction based resource allocation (PBRA), that is presented in this chapter and relies on concepts of call traffic load prediction, network cooperation, convex optimization, and decomposition theory. We first introduce the necessary modifications to the system model presented in Chap. 2 to account for the system dynamics in terms of the stochastic user mobility and call traffic models.

3.2 System Model

3.2.1 Wireless Access Networks

Consider the geographical region given in Fig. 3.1, where a set of networks $\mathcal{N} = \{1, 2, \ldots, N\}$ is available. Each network, $n \in \mathcal{N}$, is operated by a unique service provider and has a set $\mathcal{S}_n = \{1, 2, \ldots, S_n\}$ of BSs/APs. With overlapped coverage from different networks, the geographical region is partitioned into a set $\mathcal{K} = \{1, 2, \ldots, K\}$ of service areas. Each service area, $k \in \mathcal{K}$, is covered by a unique subset of BSs/APs. Let $\mathcal{N}_k$ denote the set of networks available at service area k, and $\mathcal{S}_{nk}$ denote the set of BSs/APs from network n covering service area k. Each network, $n \in \mathcal{N}$, has a downlink transmission capacity of C_n Mbps. An identification (ID) beacon is broadcasted by each network, $n \in \mathcal{N}$, BS/AP $s \in \mathcal{S}_n$, that is used in the MT attachment procedure [20]. Different networks are assumed to be connected through a backbone to exchange their roaming signalling information. The roaming signalling backbone is used to exchange signalling information among different networks, which is required by the PBRA algorithm.

3.2.2 Transmission Model

An MT, m, can get its required bandwidth, B_m, on the downlink from all wireless access networks available at its location using its multi-homing capability. Let $\mathcal{M}$ denote the set of MTs available in the geographical region, and $\mathcal{M}_{ns}$ denote the set of MTs which lie in the coverage area of the sth BS/AP of the nth network. Each MT has its own home network but can also get service from other available networks. An MT, m, which uses its own home network is referred to as network subscriber, while an MT which uses a network other than its home network is referred to as

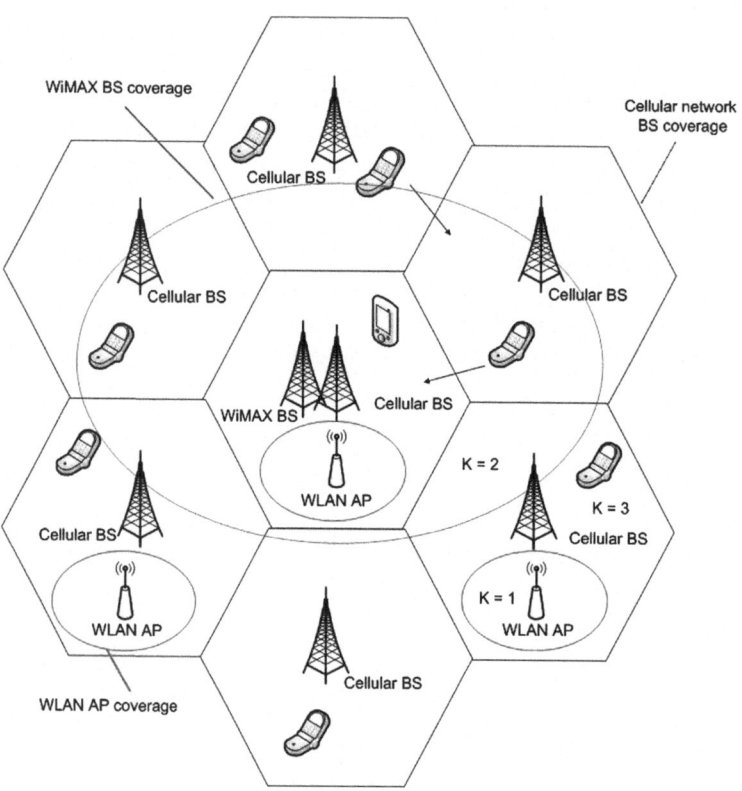

Fig. 3.1 The network coverage areas

network user. A priority parameter p_{nms} is used to differentiate the radio resource allocation to network subscribers and network users, where $p_{nms} = 1$ for network subscribers and $p_{nms} \in [0, 1)$ for network users. The allocated bandwidth in the downlink from network n to an MT m through BS/AP s is denoted by b_{nms}, with $n \in \mathcal{N}, m \in \mathcal{M}_{ns}$ and $s \in \mathcal{S}_n$. Let $B = [b_{nms}]$ be a matrix of bandwidth allocated from network n through BS/AP s to MT $m \in \mathcal{M}$, and $b_{nms} = 0$ if MT $m \notin \mathcal{M}_{ns}$. Although the presented algorithm in this chapter studies radio resource allocation on the downlink, it can be employed for radio resource allocation on the uplink.

3.2.3 Service Traffic Models

In this chapter and the next one, we only focus on VBR calls as they are more challenging to support in a dynamic system due to the requirement of providing the call with a bandwidth allocation that is as close as possible to the maximum required bandwidth. The extension of the proposed resource allocation algorithm (PBRA) is straight forward, to include CBR calls. Hence, we consider video service applications such as on-demand streaming. An MT, m, with a video call is supported by a VBR service that is allocated a bandwidth $B_m \in [B_m^{\min}, B_m^{\max}]$ [37], where $B_m^{\min}$ is the minimum required bandwidth for the video call. With more allocated bandwidth to a video call, higher perceived video quality can be experienced on the MT. A maximum bandwidth $B_m^{\max}$ can be allocated to a video call, which is enforced to incorporate the MTs technical limitations [37].

There exists a set $\mathcal{L}$ of service classes, $\mathcal{L} = \{1, 2, \ldots, L\}$. Each service class, $l \in \mathcal{L}$, has unique $B_l^{\min}$ and $B_l^{\max}$ values. For subscribers of a given network n, let M_{lk}^n denote the number of existing calls of service class l in service area k. It is assumed that there exists sufficient capacity through the available BSs/APs in the geographical region to satisfy a target call blocking probability for each service class l in each service area k for network n subscribers. Let $\widetilde{C}_{lk}^n$ denote the maximum number of calls of each service class l which can be supported in each service area k for network n subscribers, given the transmission capacities of available BSs/APs. A capacity analysis similar to the one in [60] can be used to determine this maximum number of calls. A call admission control procedure is assumed to be in place in order to ensure that $M_{lk}^n \leq \widetilde{C}_{lk}^n$, and hence feasible resource allocation solutions exist.

A Poisson process is used to model video call arrivals, which is a widely adopted assumption [60]. In particular, a Poisson process with parameter υ_{lk} is used to model the arrival process of both new and handoff video calls from service class l to service area k. Following the statistics of on-demand video streaming [34, 61], the video call duration is very likely to be heavy-tailed. The 'mice-elephants' phenomenon is a very important feature of heavy-tailedness [5]. This implies that, with respect to the video call duration, most video calls have quite short duration, while a small fraction of video calls have an extremely large duration. Yet, performance analysis is complex with heavy-tailed distributions. Hence, it is proposed in [18], for effective analysis, to fit a large class of heavy-tailed distributions with hyper-exponential distributions. For simplicity, a two-stage hyper-exponential distribution is used to model the video call duration. Thus, a video call of MT m that belongs to class l has a call duration, T_c^l, with a mean $\bar{T}_c^l$ and a probability density function (PDF), which is given by

$$f_{T_c^l}(t) = \frac{a_l}{a_l + 1} \cdot \frac{a_l}{\bar{T}_c^l} \cdot e^{-\frac{a_l}{\bar{T}_c^l} t} + \frac{1}{a_l + 1} \cdot \frac{1}{a_l \bar{T}_c^l} \cdot e^{-\frac{1}{a_l \bar{T}_c^l} t}, \qquad a_l \geq 1, t \geq 0. \quad (3.1)$$

The parameter a_l in (3.1) can characterize the mice-elephant feature. A large fraction of calls $\frac{a_l}{a_l+1}$ has a duration with mean time $\frac{\bar{T}_c^l}{a_l}$, while the other fraction $\frac{1}{a_l+1}$ has a duration with mean time $a_l \bar{T}_c^l$.

3.2.4 Mobility Models and Channel Holding Time

User residence time is used to characterize the user mobility within a given service area $k \in \mathcal{K}$, and is assumed to follow an exponential distribution. The PDF of the user residence time T_r^k, with mean $\bar{T}_r^k$, in service area $k \in \mathcal{K}$, is given by

$$f_{T_r^k}(t) = \frac{1}{\bar{T}_r^k} e^{-\frac{t}{\bar{T}_r^k}}, \quad t \geq 0. \tag{3.2}$$

In a given service area $k \in \mathcal{K}$, the channel holding time is given by $T_h^{lk} = \min(T_c^l, T_r^k)$, where T_c^l and T_r^k are independent of each other. Then,

$$Pr\{\min(T_c^l, T_r^k) > t\} = Pr\{T_c^l > t, T_r^k > t\} = Pr\{T_c^l > t\} \cdot Pr\{T_r^k > t\}. \tag{3.3}$$

This results in a channel holding time with a PDF given by

$$f_{T_h^{lk}}(t) = f_{T_c^l}(t)[1 - F_{T_r^k}(t)] + f_{T_r^k}(t)[1 - F_{T_c^l}(t)], \qquad t \geq 0 \tag{3.4}$$

where $F_{T_c^l}(t)$ and $F_{T_r^k}(t)$ are the cumulative distribution functions (CDFs) for the call duration and user residence time respectively. From (3.1) and (3.2), we have

$$f_{T_h^{lk}}(t) = \frac{a_l}{a_l + 1} \cdot \left(\frac{1}{\bar{T}_r^k} + \frac{a_l}{\bar{T}_c^l} \right) \cdot e^{-\left(\frac{1}{\bar{T}_r^k} + \frac{a_l}{\bar{T}_c^l} \right)t}$$

$$+ \frac{1}{a_l + 1} \cdot \left(\frac{1}{\bar{T}_r^k} + \frac{1}{a_l \bar{T}_c^l} \right) \cdot e^{-\left(\frac{1}{\bar{T}_r^k} + \frac{1}{a_l \bar{T}_c^l} \right)t}, \qquad t \geq 0. \tag{3.5}$$

3.3 Constant Price Resource Allocation (CPRA)

The radio resource allocation problem for MTs with multi-homing capabilities in a heterogeneous wireless access medium is given in Chap. 2 by

$$\max_{B \geq 0} \quad \sum_{n=1}^{N} \sum_{s=1}^{S_n} \sum_{m \in \mathcal{M}_{ns}} \ln(1 + \eta_1 b_{nms}) - \eta_2(1 - p_{nms})b_{nms}$$

$$s.t. \quad \sum_{m \in \mathcal{M}_{ns}} b_{nms} \leq C_{ns}, \quad \forall n \in \mathcal{N}, s \in \mathcal{S}_n \tag{3.6}$$

$$B_m^{\min} \leq \sum_{n=1}^{N} \sum_{s=1}^{S_n} b_{nms} \leq B_m^{\max}, \qquad \forall m \in \mathcal{M}$$

where η_1 and η_2 are used for scalability.

For efficient decentralized radio resource allocation in a dynamic network environment, one strategy is to avoid solving problem (3.6) for every call arrival to or departure from any service area k. Meanwhile, our main objective is to satisfy the required resource allocation per call for a target call blocking probability. This can be achieved through employing fixed link access price values for radio resource allocation at different BSs/APs independent of call arrivals and departures. Using time-invariant BS/AP link access price values, the corresponding radio resource allocation is referred to as constant price resource allocation (CPRA) [24]. The CPRA works in two phases, namely setup phase and operation phase. The setup phase takes place only once at the initial operation time of the networks, while the operation phase takes place every time a new MT joins the networks.

3.3.1 The Setup Phase

The main objective of this phase is to determine the fixed BS/AP link access price values that will be used during the operation phase. These are based on steady-state statistics of call traffic and user mobility so as to achieve satisfactory performance in terms of the allocated resources per call and call blocking probability in the operation phase.

Consider the geographical region shown in Fig. 3.1. In the setup phase, let the number of calls of each service class l in each service area k for subscribers of a given network n, M_{lk}^n, equals to a target value $\widehat{M}_{lk}^n$. The corresponding optimal link access price value for each BS/AP in the geographical region can be determined using the DORA algorithm with $\widehat{M}_{lk}^n$ values for all $n \in \mathcal{N}$ subscribers of different networks and $\forall l \in \mathcal{L}, k \in \mathcal{K}$. The radio resources of all networks will be distributed exactly over $\widehat{M}_{lk}^n$ calls $\forall n \in \mathcal{N}, l \in \mathcal{L}, k \in \mathcal{K}$, if we employ these BS/AP link access price values for resource allocation in the operation phase. Thus, for subscribers of a given network n, when $M_{lk}^n = \widehat{M}_{lk}^n$ in the operation phase, any incoming call from a network n subscriber with service class l to service area k will be blocked. This means that the choice of the target value $\widehat{M}_{lk}^n$ for all networks' subscribers and $\forall l \in \mathcal{L}, k \in \mathcal{K}$, and in turn the corresponding BS/AP link access price λ_{ns} $\forall n \in \mathcal{N}, s \in \mathcal{S}_n$, in the setup phase determines the geographical region overall performance in terms of the allocated resources per call and the call blocking probability in the operation phase. Hence, the value of $\widehat{M}_{lk}^n$ should be properly chosen to achieve target performance in the resource allocation. For a dynamic system, M_{lk}^n is a random variable. Alternatively, we can represent $\widehat{M}_{lk}^n$ by a design parameter ϵ_{lk}^n using the probability distribution of M_{lk}^n for subscribers of every network n and $\forall l \in \mathcal{L}, k \in \mathcal{K}$, such that

$$Pr(M_{lk}^n > \widehat{M}_{lk}^n) \le \epsilon_{lk}^n, \qquad \forall n \in \mathcal{N}, l \in \mathcal{L}, k \in \mathcal{K} \tag{3.7}$$

where $\epsilon_{lk}^n \in [0, 1]$. It is evident that the value of $\widehat{M}_{lk}^n$ depends on both ϵ_{lk}^n and the distribution of M_{lk}^n. Indeed, from (3.7), ϵ_{lk}^n gives an upper bound of the call blocking probability for subscribers of a given network n with service class l in service area k when $\widehat{M}_{lk}^n \leq \widetilde{C}_{lk}^n$. Otherwise, let $\widehat{M}_{lk}^n = \widetilde{C}_{lk}^n$, and both the optimal solution of (3.6) and the CPRA result in the same call blocking performance. Hence, the value of $\widehat{M}_{lk}^n$ can be chosen based on the requirement on call blocking probability for a given network n with service class l in service area k.

As call arrivals of service class l to service area k follow a Poisson process, the channel holding time follows a general distribution, and all calls are served simultaneously without queuing, an $M/G/\infty$ model [21] can be used to determine $\widehat{M}_{lk}^n$ for network n subscribers and $\forall l \in \mathcal{L}, k \in \mathcal{K}$ in the setup phase, using the steady-state call traffic and user mobility statistics. Let v_{lk}^n denote the arrival rate of new and handoff calls from network n subscribers with service class l in service area k. A BS/AP in k can determine v_{lk}^n by counting the number of new and handoff call arrivals from network n subscribers with service class l to service area k and divide it by the total elapsed time. Then, the number of calls for network n subscribers with service class l that are simultaneously present in service area k, M_{lk}^n, follows a Poisson distribution with mean $r_{lkn} = v_{lk}^n . E[T_h^{lk}]$ [21], where $E[T_h^{lk}]$ denotes the average channel holding time of service class l in service area k and can be calculated using (3.5) as

$$E[T_h^{lk}] = \frac{a_l}{a_l + 1} \cdot \frac{1}{\frac{1}{T_r^k} + \frac{a_l}{T_c^l}} + \frac{1}{a_l + 1} \cdot \frac{1}{\frac{1}{T_r^k} + \frac{1}{a_l T_c^l}}, \qquad \forall l \in \mathcal{L}, k \in \mathcal{K}. \qquad (3.8)$$

Hence, from (3.7), $\widehat{M}_{lk}^n$ is the minimum integer which satisfies [21]

$$\sum_{i=0}^{\widehat{M}_{lk}^n} \frac{r_{lkn}^i e^{-r_{lkn}}}{i!} \geq (1 - \epsilon_{lk}^n), \qquad \forall n \in \mathcal{N}, l \in \mathcal{L}, k \in \mathcal{K}. \qquad (3.9)$$

For a given ϵ_{lk}^n, using $\widehat{M}_{lk}^n \, \forall n \in \mathcal{N}, l \in \mathcal{L}, k \in \mathcal{K}$, problem (3.6) can be solved using the DORA algorithm in order to find the corresponding optimal link access price values $\widehat{\lambda}_{ns} \, \forall n \in \mathcal{N}, s \in \mathcal{S}_n$.

3.3.2 The Operation Phase

The main objective of this phase is to perform the bandwidth allocation process for each user joining the networks based on the following four steps.

Step 1: Each network BS/AP in the geographical region fixes its link access price value to the value calculated in the setup phase, $\widehat{\lambda}_{ns}$, independent of call arrivals and departures. This fixed value, $\widehat{\lambda}_{ns}$, is broadcasted by each network $n \in \mathcal{N}$ BS/AP $s \in \mathcal{S}_n$ via its ID beacon.

Table 3.1 Calculation of bandwidth share from each available network BS/AP at MT m

1: **Input:** $\widehat{\lambda}_{ns}$ $\forall n \in \mathcal{N}_k, s \in \mathcal{S}_{nk}, B_m, m \in \mathcal{M}$;

2: **Initialization:** $\mu_m^{(1)}(1) \geq 0; \mu_m^{(2)}(1) \geq 0$;

3: **for** $i = 1 : I$ **do**

4:     **for** $n \in \mathcal{N}_k$ **do**

5:         **for** $s \in \mathcal{S}_{nk}$ **do**

6:             $b_{nms}(i) = [(\frac{\eta_1}{\widehat{\lambda}_{ns}+(\mu_m^{(1)}(i)-\mu_m^{(2)}(i))+\eta_2(1-p_{nms})} - 1)/\eta_1]^+$;

7:         **end for**

8:     **end for**

9:     $\mu_m^{(1)}(i+1) = [\mu_m^{(1)}(i) - \alpha_1(B_m^{\max} - \sum_{n=1}^{N}\sum_{s=1}^{S_n} b_{nms}(i))]^+$;

10:   $\mu_m^{(2)}(i+1) = [\mu_m^{(2)}(i) - \alpha_2(\sum_{n=1}^{N}\sum_{s=1}^{S_n} b_{nms}(i) - B_m^{\min})]^+$;

11: **end for**

12: **Output:** The required b_{nms} $\forall n \in \mathcal{N}_k, s \in \mathcal{S}_{nk}$.

Step 2: An incoming MT listens to the link access price values of the BSs/APs available at its location through its multiple radio interfaces.

Step 3: The link access price values are then used by the MTs in order to solve for the bandwidth share from each network BS/AP such that the total amount of allocated resources from all BSs/APs satisfies the call required bandwidth. This can be calculated at MT, m, with service class l in service area k, using the algorithm in Table 3.1, which is based on the DORA algorithm.

Step 4: MT, m, asks BS/AP s of network n, $\forall n \in \mathcal{N}_k, s \in \mathcal{S}_{nk}$, for the calculated bandwidth share b_{nms}. The BS/AP performs the required bandwidth allocation if it has sufficient resources. The MT call is blocked if the call total required bandwidth is not satisfied by the total allocated radio resources.

In the CPRA, no resource reallocations to existing calls are required since the BS/AP link access price values are independent of call arrivals to and departures from different service areas. Moreover, the required I iterations to reach the desired resource allocations from all BSs/APs to satisfy the call total required bandwidth is solved locally at each MT. Hence, no information exchange is required between the MTs and the BSs/APs for every iteration as in the DORA algorithm. Thus, almost no signalling overhead is required in the CPRA in order to reach the required bandwidth from each BS/AP.[1] The convergence of the CPRA follows the convergence of the DORA algorithm which is given in Chap. 2. However, unlike the DORA algorithm, the CPRA provides a sub-optimal solution to problem (3.6) since the link access price value is not updated with every call arrival and departure.

In the CPRA, a low call blocking probability can be obtained in the operation phase using a small value of ϵ_{lk}^n. However, this corresponds to a large $\widehat{M}_{lk}^n$ value. This results in a large BS/AP link access price values, which leads to a low amount of resource allocation per call in the operation phase. On the other hand, a large ϵ_{lk}^n value results in a high call blocking probability and a large amount of resource

---

[1] This is apart from the required overhead in broadcasting the fixed link access price value $\widehat{\lambda}_{ns}$ by every BS/AP on its ID beacon. However, the contribution of broadcasting this value to the overhead is negligible.

allocation per call in the operation phase. As a result, the value of ϵ_{lk}^n should be chosen so as to balance the trade-off between the allocated resources per call and the call blocking probability.

Using an appropriate choice of ϵ_{lk}^n, the CPRA with its setup and operation phases can allocate radio resources for a target call blocking probability in the decentralized network architecture with dynamic call arrivals and departures.

3.4 Prediction Based Resource Allocation (PBRA)

The CPRA is performed based on $\widehat{M}_{lk}^n$ which is calculated according to the steady-state (long-term) call traffic and user mobility statistics. However, in a dynamic environment, with call arrivals and departures, M_{lk}^n can deviate from $\widehat{M}_{lk}^n$ for some time. Yet, the allocated resources in the operation phase do not adapt to short-term dynamics in the call traffic load. Hence, even if there exist sufficient resources in the BSs/APs that can be used to improve a video call quality, the call can be allocated only its minimum required bandwidth. In CPRA, these unutilized extra resources (at a low call traffic load) are actually reserved for possible incoming calls so as to satisfy the target call blocking probability. A resource allocation adaptive to a short-term call traffic load (via resource re-allocation to the calls in service) can help to provide a better service quality compromise between the existing calls (in terms of the amount of allocated resources to each call) and the potential incoming calls (in terms of the call blocking probability). Towards this end, in the following, we propose to update $\widehat{M}_{lk}^n$ $\forall n \in \mathcal{N}, l \in \mathcal{L}, k \in \mathcal{K}$ in the operation phase periodically with period τ, and hence update the corresponding BS/AP link access price values, based on the instantaneous M_{lk}^n value at time t, $M_{lk}^n(t)$. We refer to the corresponding resource allocation as prediction based resource allocation (PBRA) [24].

Let the time be partitioned into a set of periods, $\mathcal{T}$, of constant duration τ, $\mathcal{T} = \{T_1, T_2, \ldots, T_j, \ldots\}$. Let t_j denotes the beginning of each period T_j. A time vector of arrival events for calls of network n subscribers with service class l in service area k during period T_j is denoted by $\boldsymbol{T}_{lkn}^j$. The PBRA algorithm is carried out in the following six steps.

Step 1: Given a new call arrival at time instant $t_a^j \in \boldsymbol{T}_{lkn}^j$, $a = \{1, 2, \ldots, |\boldsymbol{T}_{lkn}^j|\}$, in period T_j, the number of calls of network n subscribers with service class l in service area k at the time instant, $M_{lk}^n(t_a^j)$, is used by the BSs/APs in this service area to probabilistically predict the number of calls at time instant $t_a^j + \tau$ in the next time period T_{j+1}. Hence, τ is referred to as the prediction duration. The predicted number, $\widetilde{M}_{lk}^n(t_a^j + \tau)$, should satisfy

$$Pr(M_{lk}^n(t_a^j + \tau) > \widetilde{M}_{lk}^n(t_a^j + \tau)|M_{lk}^n(t_a^j)) \le \epsilon_{lk}^n, \qquad \forall n \in \mathcal{N}, l \in \mathcal{L}, k \in \mathcal{K}. \quad (3.10)$$

In order to determine $\widetilde{M}_{lk}^n(t_a^j + \tau)$, the conditional probability mass function (PMF) of $M_{lk}^n(t_a^j + \tau)$ given $M_{lk}^n(t_a^j)$, $P_{M_{lk}^n(t_a^j+\tau)|M_{lk}^n(t_a^j)}(i)$, is calculated using the transient distribution of the $M/G/\infty$ model [40]. First, we present the following definitions under the assumption of stationary call arrival and departure processes:

- p_τ^{lkn}—The probability that a call of network n subscribers with service class l which is in service area k at time t_a^j is still present in the same service area at time $t_a^j + \tau$;
- q_τ^{lkn}—The probability that a call of network n subscribers with service class l that arrives in service area k during $(t_a^j, t_a^j + \tau]$ is still present in the same service area at time $t_a^j + \tau$;
- $X_B(\kappa, \alpha)$—A binomial random variable with parameters κ and α;
- $X_P(\alpha)$—A Poisson random variable with mean α.

Using $M_{lk}^n(t_a^j)$, we have [40]

$$M_{lk}^n(t_a^j + \tau) =_d X_B(M_{lk}^n(t_a^j), p_\tau^{lkn}) + X_P(\upsilon_{lk}^n \tau q_\tau^{lkn}) \qquad (3.11)$$

where $=_d$ denotes equality in distribution. The probabilities p_τ^{lkn} and q_τ^{lkn} are defined as [40]

$$p_\tau^{lkn} = \frac{1}{E[T_h^{lk}]} \int_\tau^\infty Pr(T_h^{lk} > s)ds = \frac{1}{E[T_h^{lk}]} \int_\tau^\infty (1 - F_{T_h^{lk}}(s))ds. \qquad (3.12)$$

$$q_\tau^{lkn} = \int_0^\tau \frac{1}{\tau} Pr(T_h^{lk} > s)ds = \int_0^\tau \frac{1}{\tau}(1 - F_{T_h^{lk}}(s))ds = \frac{E[T_h^{lk}]}{\tau}(1 - p_\tau^{lkn}) \qquad (3.13)$$

where $F_{T_h^{lk}}(s) = \int_0^s f_{T_h^{lk}}(t)dt$ is the CDF of T_h^{lk}. The conditional PMF, $P_{M_{lk}(t_a^j+\tau)|\ M_{lk}(t_a^j)}(i)$, can be calculated using (3.11)–(3.13). Hence, $\widetilde{M}_{lk}(t_a^j + \tau)$ can be calculated using (3.10) as the minimum integer satisfying

$$\sum_{i=0}^{\widetilde{M}_{lk}^n(t_a^j+\tau)} P_{M_{lk}^n(t_a^j+\tau)|M_{lk}^n(t_a^j)}(i) \geq (1 - \epsilon_{lk}^n), \qquad \forall n \in \mathcal{N}, l \in \mathcal{L}, k \in \mathcal{K}. \qquad (3.14)$$

Step 2: Each BS/AP in the geographical region records the predicted values of $\widetilde{M}_{lk}^n(t_a^j + \tau), \forall n \in \mathcal{N}, l \in \mathcal{L}, k \in \mathcal{K}$ for $a = \{1, 2, \ldots, |\mathcal{T}_{lkn}^j|\}$ in a vector $\mathcal{M}_{lkn}^{j+1}$.

Step 3: For the subscribers of each network $n \in \mathcal{N}$, the maximum predicted number of calls from each service class $l \in \mathcal{L}$ in each service area $k \in \mathcal{K}$ during T_{j+1}, $\widetilde{M}_{lk}^n(T_{j+1})$, is calculated at t_{j+1} from $\mathcal{M}_{lkn}^{j+1}$. Hence $\widetilde{M}_{lk}^n(T_{j+1}) = \max(\mathcal{M}_{lkn}^{j+1})$ if it is less than or equal to $\widetilde{C}_{lk}^n$, otherwise $\widetilde{M}_{lk}^n(T_{j+1}) = \widetilde{C}_{lk}^n$. This ensures that for $\widetilde{M}_{lk}(T_{j+1}) \leq \widetilde{C}_{lk}$, we have

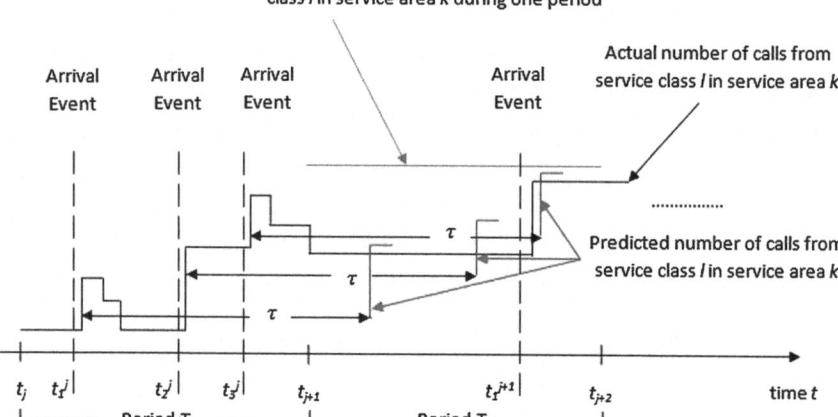

Fig. 3.2 Illustration of PBRA Events

$$Pr(M_{lk}^n(t_a^{j+1}) > \widetilde{M}_{lk}^n(T_{j+1})) \le \epsilon_{lk}^n,$$

$$\forall n \in \mathcal{N}, l \in \mathcal{L}, k \in \mathcal{K}, a \in \{1, 2, \ldots, |\boldsymbol{T}_{lkn}^{j+1}|\}. \tag{3.15}$$

Step 4: Through cooperative networking, different BSs/APs in the geographical region exchange their information regarding $\widetilde{M}_{lk}^n(T_{j+1})$ $\forall n \in \mathcal{N}, l \in \mathcal{L}, k \in \mathcal{K}$. Problem (3.6) can be solved at each BS/AP to update its link access price value which is fixed over T_{j+1}, independent of call arrivals to and departures from different service areas, and is broadcasted on the BS/AP ID beacon.

Figure 3.2 illustrates the call arrival times, the actual and predicted numbers of calls for network n subscribers with service class l in service area k associated with the steps 1–4.

Step 5: During T_{j+1}, each MT in the geographical region, including both incoming and already existing ones, uses the broadcasted BS/AP link access price values received at its location during this period to determine and ask for a bandwidth share from each available BS/AP. This is achieved following steps 2–4 in the CPRA.[2]

Step 6: Each MT reports to the BSs/APs available at its location its service class, home network, and a list of the BS/AP IDs that the MT can receive. BSs/APs of different networks use this information so as to predict $\widetilde{M}_{lk}^n(T_{j+2})$, $\forall n \in \mathcal{N}, l \in \mathcal{L}, k \in \mathcal{K}$, during the next period T_{j+2} to update their link access price values at time t_{j+2}.

While the CPRA uses the target $\widehat{M}_{lk}^n$ value from the setup phase based on steady-state (long-term) statistics to perform the radio resource allocation in the operation phase, the PBRA updates the target value by $\widetilde{M}_{lk}^n(T_j)$ every period $T_j, j = \{1, 2, \ldots\}$, using the current number of calls in service. Using this extra information, the PBRA

---

[2] In Table 3.1, $\widehat{\lambda}_{ns}$ is replaced by the updated link access price value the MT receives during T_{j+1}.

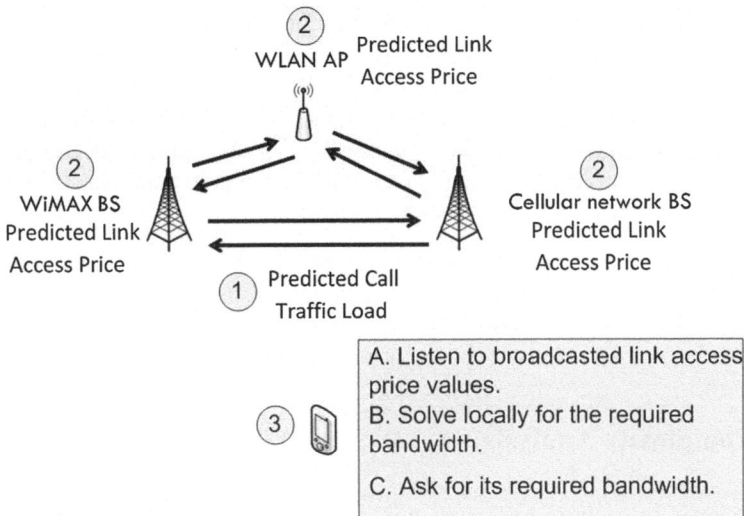

Fig. 3.3 The PBRA procedure

can make a better prediction of the call traffic load carried in the geographical region in a short-term, and hence an improved radio resource allocation is expected over the CPRA. The PBRA algorithm provides an improved sub-optimal solution to problem (3.6) as compared to the CPRA algorithm. The convergence of the PBRA algorithm to this sub-optimal solution follows the convergence of the DORA algorithm which is given in Chap. 2. As the BS/AP link access price values during period T_j are based on $\widetilde{M}_{lk}^n(T_j)$, the BSs/APs allocate their available resources exactly among $\widetilde{M}_{lk}^n(T_j)$ calls during period T_j. Thus, following the definitions in (3.10) and (3.15) and using the same argument of CPRA, ϵ_{lk}^n serves as an upper bound on the call blocking probability for $\widetilde{M}_{lk}^n(T_j) \leq \widetilde{C}_{lk}^n$.

The PBRA procedure is illustrated in Fig. 3.3. The main differences between the DORA and PBRA operations are summarized in the following:

1. In the DORA algorithm, the link access price values for different BSs/APs are updated with every call arrival to or departure from any service area. This requires resource reallocations for all existing MTs, which results in high signalling overhead. On the other hand, the PBRA updates the BSs/APs link access price values every τ, fix and broadcast them during τ. This can significantly reduce the amount of signalling overhead over the air interface, and is achieved through short-term call traffic load prediction and network cooperation. Specifically, cooperative networking allows different networks to exchange the necessary information required so as to enable each BS/AP to calculate and broadcast the predicted link access price value for the next τ duration.

2. In the DORA algorithm, each MT plays an active role in the resource allocation operation by coordinating different BSs/APs resource allocations so as to satisfy

the call total required bandwidth. While in the PBRA, the MT active role is to calculate the required bandwidth share from each BS/AP to satisfy the call total required bandwidth. Hence, in the PBRA, all the necessary information for the calculations are made locally available to the MT, unlike the DORA algorithm, which again significantly reduces the amount of signalling overhead required over the air interface in order to determine the required bandwidth share from each network BS/AP.

These differences are made clear by comparing Figs. 2.3 and 3.3. In the following section, we present a complexity analysis for the DORA implementation in a dynamic system, the CPRA, and the PBRA.

3.5 Complexity Analysis

The complexity analysis in this section examines both signalling overhead and processing time complexity for the DORA implementation in a dynamic environment, the CPRA, and the PBRA.

3.5.1 Signalling Overhead

In order to implement the DORA algorithm in a dynamic system, information signalling needs to be exchanged between all existing MTs and BSs/APs with every call arrival to and departure from any service area in order to reach the optimal resource allocation. This signalling overhead is a function of the call arrival and departure rates, the number of existing calls in different service areas, and the required number of iterations I for the DORA algorithm to converge to the optimal resource allocation. Denote χ_a and χ_d as the average number of call arrivals and departures over a period, respectively. Hence, for the DORA implementation in a dynamic system, the signalling overhead on the air interface scales as $O(\chi_a + \chi_d)$ over the period. As a result, for high call arrival/departure rates, a high signalling overhead is expected. On the other hand, for the CPRA and PBRA, the link access price values for different BSs/APs are independent of call arrivals and departures. Thus, in order to reach the required resource allocation, their signalling overhead on the air interface scales as $O(1)$. As a result, the signalling overhead for the CPRA and the PBRA scales well with the call arrival and departure rates, as compared with the DORA implementation in a dynamic system.

3.5.2 Processing Time

In the DORA algorithm, MTs and BSs/APs exchange signalling information for I iterations in order to reach an optimal resource allocation. Let σ denote the total amount of time required for the signalling exchange completion for the I iterations. The signalling exchange for I iterations should take place with every call arrival to or departure from any service area. Then, the time duration between two successive execution of the I-iteration signalling exchange is expressed as $\delta = \min$(call inter-arrival time, call departure time). Since the call arrivals follow a Poisson process with parameter υ_{lk}, the call inter-arrival time follows an exponential distribution with PDF $f_{T_a^{lk}}(t)$. The channel holding time gives the call departure time, which follows a hyper-exponential distribution with PDF $f_{T_h^l}(t)$. Using the same analysis as given in (3.3)–(3.4), the PDF of δ, $f_\delta(t)$, is expressed as

$$f_\delta(t) = \frac{a_l}{a_l + 1} \cdot \left(\frac{1}{\bar{T}_r^k} + \frac{a_l}{\bar{T}_c^l} + \upsilon_{lk} \right) \cdot e^{-\left(\frac{1}{\bar{T}_r^k} + \frac{a_l}{\bar{T}_c^l} + \upsilon_{lk} \right)t}$$

$$+ \frac{1}{a_l + 1} \cdot \left(\frac{1}{\bar{T}_r^k} + \frac{1}{a_l \bar{T}_c^l} + \upsilon_{lk} \right) \cdot e^{-\left(\frac{1}{\bar{T}_r^k} + \frac{1}{a_l \bar{T}_c^l} + \upsilon_{lk} \right)t}, \qquad t \geq 0. \quad (3.16)$$

Using (3.16), the average of δ is given by

$$\bar{\delta} = \frac{a_l}{a_l + 1} \cdot \frac{1}{\frac{1}{\bar{T}_r^k} + \frac{a_l}{\bar{T}_c^l} + \upsilon_{lk}} + \frac{1}{a_l + 1} \cdot \frac{1}{\frac{1}{\bar{T}_r^k} + \frac{1}{a_l \bar{T}_c^l} + \upsilon_{lk}}. \quad (3.17)$$

It is clear from (3.17) that the DORA algorithm processing time does not scale with the call arrival and departure rates, since $\bar{\delta}$ is inversely proportional to them. As $\bar{\delta}$ decreases with increasing arrival and/or departure rates while σ increases with increasing arrival rates (this is especially true in a case that a contention based medium access control network is among the available networks), $\bar{\delta}$ can be smaller than σ. Hence, the DORA algorithm does not converge to an optimal allocation whenever $\bar{\delta}$ is smaller than σ. On the other hand, for the CPRA and the PBRA, the I iterations are solved locally at the MTs and no signalling information is exchanged for each iteration, unlike the DORA algorithm. As a result, both the CPRA and the PBRA reach the required bandwidth allocation from each BS/AP independent of the call arrival and departure rates.

The CPRA and the PBRA require that the BS/AP link access price values to be broadcasted by each BS/AP on its ID beacon. Furthermore, the PBRA requires an exchange of the predicted call traffic load among different BSs/APs with overlapped coverage every τ. However, unlike the DORA algorithm, this signalling exchange does not take place on the air interface, but is executed over the signalling backbone connecting different networks.

Since the link access price value for different BSs/APs are updated every τ, the choice of the τ duration should reflect some change in the call traffic load in the geographical region. Hence, as a guideline, the time duration τ can be chosen such that the probability $Pr[\delta < \tau]$ is less than a small threshold γ.

3.6 Simulation Results and Discussion

In this section, we present simulation results for the radio resource allocation in a heterogeneous wireless access medium for MTs with multi-homing capabilities, using the PBRA algorithm as compared to problem (3.6) exact solution and the CPRA. Consider the geographical region of Fig. 3.4. A single VBR service class ($l = 1$) is considered and we study the performance of the PBRA algorithm in the service area ($k = 1$) which is covered by all three networks, in terms of the allocated resources per call and the call blocking probability. For simplicity, it is assumed that only subscribers of one network are present, and all networks treat them in the same manner (i.e. $p_{nms} = 1$ from all networks). The transmission capacity allocated from network n BS/AP to the service area under consideration is given by $C_1 = 4$ Mbps, $C_2 = 0.656$ Mbps, $C_3 = 2$ Mbps. A total of 26 VBR calls with required bandwidth allocation $B_m \in [0.256, 0.512]$ Mbps for MTs with multi-homing capabilities can be supported in the service area under consideration using the given C_n values, that is $\widetilde{C} = 26$ (indices n, l, k are dropped for simplicity). The new and handoff video call arrival process is modeled by a Poisson process with parameter υ (call/minute—indices n, l, k are dropped for simplicity). A hyper-exponential distribution is used to model the video call duration, with the PDF given in (3.1) and $a_1 = 1$. The average video call duration $\bar{T}_c = 20$ min. The user residence time in the service area under consideration follows an exponential distribution with the PDF given in (3.2) having an average time $\bar{T}_r = 15$ min [60]. The parameters η_1 and η_2 are set to 1 [57].

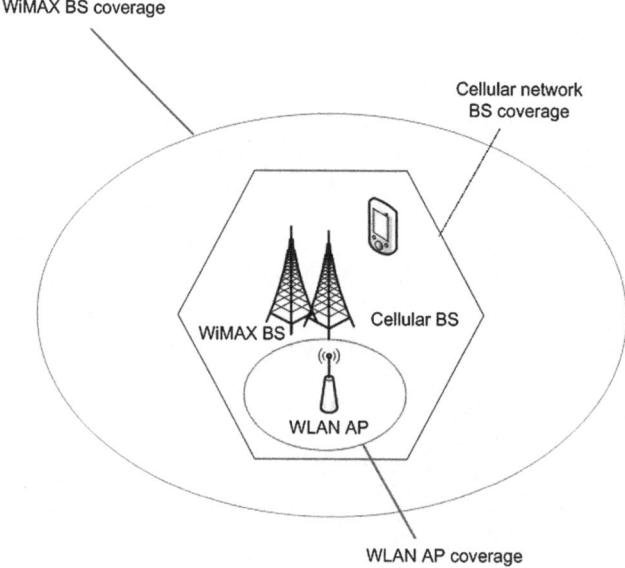

Fig. 3.4 Service areas under consideration

3.6.1 Performance Comparison

In the following, the performance of the PBRA algorithm is compared with the optimal solution of problem (3.6) in terms of the allocated resources per call and the call blocking probability. The optimal solution of (3.6) is referred to as ORAP and can be obtained using a centralized resource allocation. Although it is not appropriate for practical implementation when different networks are operated by different service providers, the ORAP is used to serve as an upper bound for the system performance in terms of the allocated resources per call and a lower bound for the system performance in terms of the call blocking probability. The CPRA is also considered in the comparison, where no update of the link access price values takes place.

Figure 3.5 shows performance comparison among the CPRA, PBRA, and ORAP versus the call arrival rate v, with $\epsilon = 1\,\%$ and $\tau = 0.25, 0.5$, and 1 min. At a low call arrival rate, the predicted number of simultaneously present calls is low, thus the estimated link access price value is low and the resource allocation amounts per call using the PBRA algorithm for the different τ values are high. On the other hand, at a high call arrival rate, the predicted number of simultaneously present calls in the system is high. For a larger τ value, less resources are allocated per call as explained in the next sub-section. The CPRA provides a lower bound of the performance in terms of the allocated resources, as it does not update the BS/AP link access price values. For the ORAP, there is no call blocking probability for a call arrival rate $v < 1.5$ call/min. All three algorithms achieve the desired upper bound for call blocking probability, ϵ, for $v \leq 1.9$ call/min. The predicted number of calls simultaneously present in the system is larger than $\widetilde{C}$ for $v > 1.9$ call/min. As a result, according to the CPRA and the PBRA, the predicted number is made equal to $\widetilde{C}$, and the algorithms achieve the same call blocking probability as the ORAP. Overall, the PBRA performance lies between CPRA and ORAP performance, as expected. By properly choosing the τ value, the PBRA algorithm can achieve a desired compromise between performance and implementation complexity.

3.6.2 Performance of the PBRA Algorithm

In the following, the performance of the PBRA algorithm is studied versus its two parameters, namely the upper bound on the call blocking probability ϵ and the prediction duration τ.

Figure 3.6 shows the performance of the PBRA algorithm in terms of the amount of resource allocation per call and call blocking probability versus ϵ, with the call arrival rate $v = 1.7$ call/min and the prediction duration $\tau = 1$ min. As ϵ increases, the PBRA algorithm accounts for the simultaneous presence of less calls in service in the next τ in its calculation of the link access price value. This results in an increase in the call blocking probability with ϵ. In general, the call blocking probability does not exceed its upper bound ϵ as shown in Fig. 3.6b. However, the resource allocation

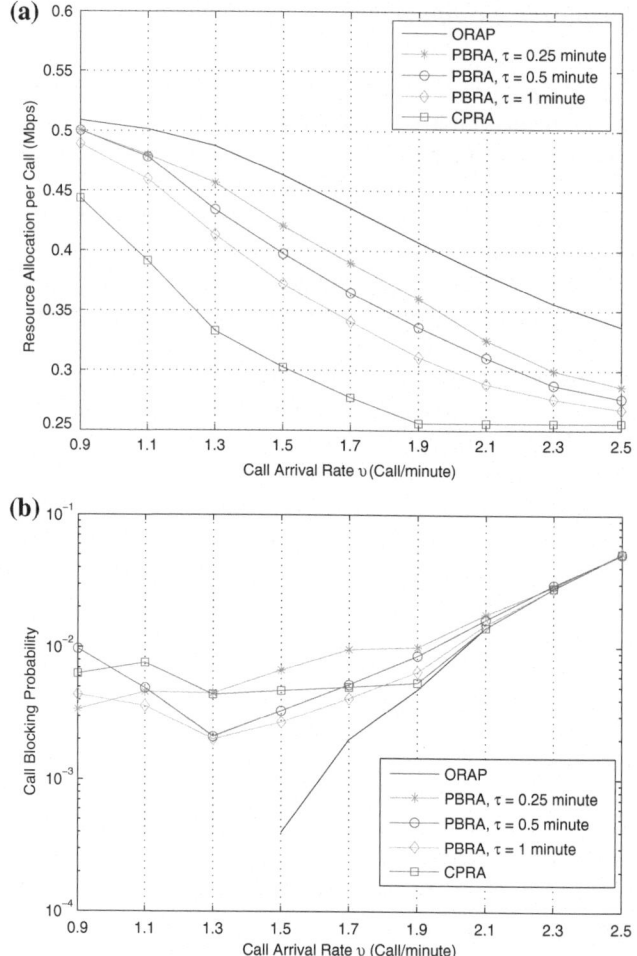

Fig. 3.5 Performance comparison: **a** resource allocation per call; **b** call blocking probability. $\epsilon = 1\%$ and $\tau = 0.25, 0.5$ and 1 min

per call is improved with ϵ, since less resources are reserved for incoming calls which will more likely be blocked. Thus, a trade-off exists between these two performance metrics.

Figure 3.7 shows the performance of the PBRA in terms of the amount of resource allocation per call and call blocking probability versus the prediction duration τ, with the call arrival rate $\upsilon = 1.7$ call/min and $\epsilon = 1\%$. With a larger prediction duration τ, the PBRA algorithm updates the BS/AP link access price less frequently and a larger number of simultaneously present calls is predicted. As a result, the amount of allocated resources per call is reduced. Again, the call blocking probability does not exceed its upper bound ϵ with the different τ values.

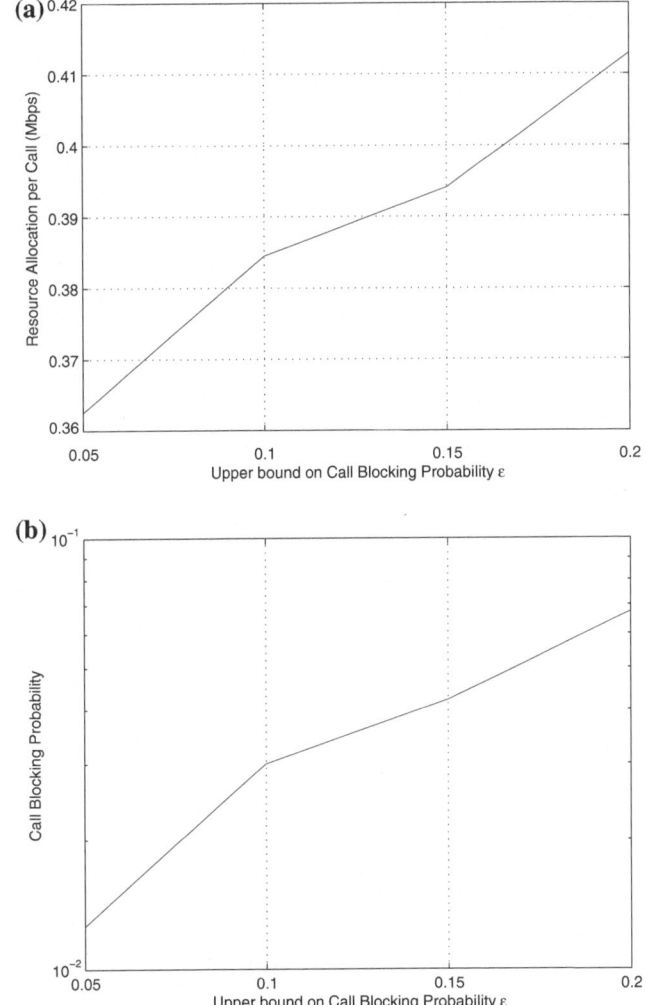

Fig. 3.6 The PBRA algorithm performance versus ϵ: **a** resource allocation per call; **b** call blocking probability. $\upsilon = 1.7$ call/min and $\tau = 1$ min

3.7 Summary

In this chapter, the limitations of the DORA algorithm in a dynamic system are discussed. A prediction based resource allocation (PBRA) algorithm is presented to address these limitations. The PBRA objective is to perform an efficient resource allocation in a dynamic system that can reduce the signalling overhead required over the air interface for resource allocation in a decentralized architecture while achieving an acceptable call blocking probability and a sufficient amount of allocated

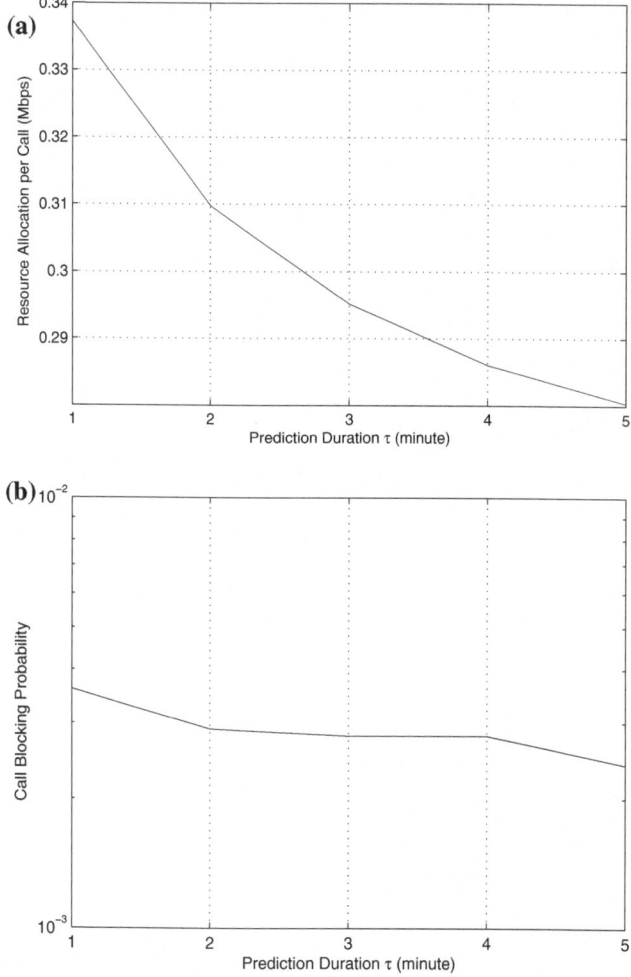

Fig. 3.7 The PBRA algorithm performance versus τ: **a** Resource allocation per call; **b** call blocking probability. $v = 1.7$ call/min and $\epsilon = 1\%$

resources per call. In order to achieve the objectives, the PBRA algorithm relies on short-term call traffic load prediction and network cooperation. There are two parameters in the PBRA algorithm, namely ϵ_{lk}^{n} and τ, that can be properly chosen to strike a balance between the desired performance in terms of the allocated resources per call and the call blocking probability, and between the performance and the implementation complexity. In the PBRA algorithm, each MT plays an active role in the resource allocation operation by requesting a bandwidth share from each available network based on the available resources at the network, such that the total allocated bandwidth from different networks satisfies the MT service requirement. However, the proposed PBRA algorithm supports only MTs with multi-homing service. It is envisioned that both single-network and multi-homing services will co-exist in the future heterogeneous wireless communication network. Hence, in the next chapter, we extend the concepts presented in Chaps. 2 and 3 to include the presence of MTs with single-network service in the networking environment and hence discuss a resource allocation algorithm that can support MTs with single-network and multi-homing services in a decentralized manner.

Chapter 4
Resource Allocation for Single-Network and Multi-Homing Services

In the future wireless communication network, it is envisioned that both single-network and multi-homing services will co-exist. Hence, it is required to develop radio resource allocation algorithms that can support both service types. In this case, the radio resource allocation mechanism is to determine the optimal network assignment for MTs with single-network service and the corresponding bandwidth allocation for MTs with single-network and multi-homing services. In this chapter, we discuss how to achieve these objectives in a decentralized network architecture with call arrivals and departures. Concepts of call traffic load prediction and cooperative networking, presented in Chap. 3, are employed to enable vertical handovers for single-network calls in a seamless manner and to satisfy multi-homing calls required bandwidth in such a decentralized network architecture.

4.1 Introduction

In Chaps. 2 and 3, we have presented a set of algorithms to support a decentralized resource allocation for MTs in a heterogeneous wireless access medium. In these algorithms, an MT plays an active role in the resource allocation operation, whether by coordinating the resource allocation from different networks or by calculating the required bandwidth share from each network and asking for this share to satisfy its total required bandwidth. However, the algorithms can support only MTs with multi-homing capabilities. It is expected that both single-network and multi-homing services will coexist in future wireless networks. Many reasons support this vision. On one hand, not all calls require high data rates that call for a multi-homing support, and hence these calls can resort to a single-network service. In addition, not all MTs are currently equipped with multi-homing capabilities, thus they can only support a single-network service. Moreover, an MT with insufficient available energy can switch from a multi-homing service to a single-network service and turn off all its radio interfaces, except for the one with the best available wireless network, in order to save energy. Hence, it is required to develop a decentralized radio resource

M. Ismail and W. Zhuang, *Cooperative Networking in a Heterogeneous Wireless Medium*, SpringerBriefs in Computer Science, DOI: 10.1007/978-1-4614-7079-3_4, © The Author(s) 2013

allocation algorithm that can support both single-network and multi-homing services. In such a decentralized architecture, an MT with single-network service should be able to select the best available wireless access network at its location and ask for its required bandwidth from this network. In addition, the MT should be able to perform a vertical handover whenever necessary so as to remain best connected. On the other hand, an MT with multi-homing service can determine the required bandwidth share from each network to satisfy its total required bandwidth. Hence, the objective of the radio resource allocation algorithm is twofold: First, to determine the optimal network assignment vector for MTs with single-network service; Second, to determine the corresponding optimal bandwidth allocation for MTs with single-network and multi-homing services. Towards this end, we first present a centralized optimal radio resource allocation (CORA) algorithm that can satisfy the aforementioned objectives. Then, based on the centralized algorithm and the concepts introduced in Chap. 3 for call traffic load prediction and network cooperation, we present a decentralized sub-optimal resource allocation (DSRA) algorithm. In the next section, we first introduce the necessary modifications to the system model presented in Chap. 3 to account for the presence of single-network calls in the networking environment.

4.2 System Model

4.2.1 Wireless Access Networks

Consider geographical region with a set, $\mathcal{N} = \{1, 2, \ldots, N\}$, of available wireless access networks. Each network, $n \in \mathcal{N}$, is operated by a unique service provider and has a set, $\mathcal{S}_n = \{1, 2, \ldots, S_n\}$, of BSs/APs in the geographical region. The BSs/APs of different networks have overlapped coverage in some areas which partitions the region into a set, $\mathcal{K} = \{1, 2, \ldots, K\}$, of service areas. Each service area, $k \in \mathcal{K}$, is covered by a unique subset of BSs/APs, as shown in Fig. 4.1. Let $\mathcal{N}_k$ denote the subset of available networks at service area k, and $\mathcal{S}_{nk}$ denote the subset of BSs/APs from network n covering service area k. The subset $\mathcal{S}_k$ denotes the BSs/APs from all networks covering service area k, with cardinality $|\mathcal{S}_k|$. Each BS/AP, $s \in \mathcal{S}_n$, has a downlink transmission capacity C_n Mbps. An identification (ID) beacon is broadcasted by each BS/AP, which is used in the MT attachment procedure [20]. It is assumed that different networks are connected through a backbone to exchange their roaming signalling information. We rely on the roaming signalling backbone in order to exchange the signalling information required by the DSRA algorithm.

4.2.2 Service Types

The set of MTs in the geographical region is denoted by $\mathcal{M}$. Let $\mathcal{M}_k$ denotes the subset of MTs in a given service area, k. Each MT has its own home network but can

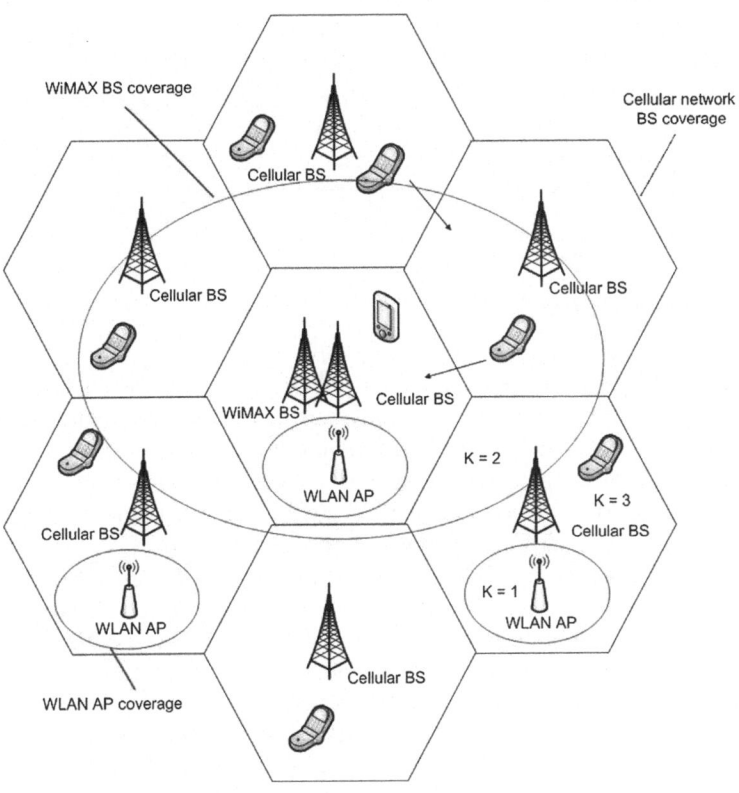

Fig. 4.1 The network coverage areas

also get service from other available networks. An MT, m, which uses its own home network is referred to as network subscriber, while an MT which uses a network other than its home network is referred to as network user. A priority parameter p_{nms} is used to differentiate the radio resource allocation to network subscribers and network users, where $p_{nms} = 1$ for network subscribers and $p_{nms} \in [0, 1]$ for network users. Two service types are considered, i.e. single-network and multi-homing services. Let $\mathcal{M}_{vk}$ denote the subset of MTs with same service type in a given service area k, where $v = 1$ for single-network service and $v = 2$ for multi-homing service. An MT with single-network service, $m \in \mathcal{M}_{1k}$ in service area k, is assigned to a single network n BS/AP $s \in \mathcal{S}_{nk}$. The network assignment criterion is based on the network offered bandwidth for the MT. Hence, the MT is assigned to a network $n \in \mathcal{N}$ BS/AP $s \in \mathcal{S}_n$ that can support it with the largest bandwidth allocation as compared to other BSs/APs in the service area. Let $A = [a_1, \ldots, a_m, \ldots, a_{|\mathcal{M}_{1k}|}]$ denote the network assignment vector in the geographical region for MTs with single-network service, where $a_m = ns$ is the assignment of MT $m \in \mathcal{M}_{1k}$ to network n BS/AP s. For instance, $a_1 = 12$ is the assignment of MT 1 to network 1 BS/AP 2.

On the other hand, an MT, $m \in \mathcal{M}_{2k}$, with multi-homing service in a given service area k, receives its required bandwidth from all BSs/APs available at its location, $s \in \mathcal{S}_k$, using its multi-homing capability. The set $\mathcal{M}_{ns}$ of MTs assigned to network n BS/AP s includes both multi-homing and single-network MTs.

4.2.3 Service Traffic Models

A video call to MT m is considered to be a VBR service that is allocated a total bandwidth of B_m in the range $[B_m^{\min}, B_m^{\max}]$, where $B_m^{\min}$ is the total minimum required bandwidth by MT m which guarantees a minimum QoS requirement for the video call, and $B_m^{\max}$ is the total maximum required bandwidth by MT m which is enforced to incorporate the MT technical limitations. The more allocated bandwidth to a video call, the higher the perceived video quality experienced on the MT.

There exists a set, $\mathcal{L}_v = \{1, 2, \ldots, L_v\}$, of service classes for each service type, v. In general, service class $\mathcal{L}_2$ for an MT with multi-homing service type requires larger bandwidth than service class $\mathcal{L}_1$ for an MT with single-network service. The total allocated bandwidth to MT m with a VBR call of service type v and service class l is B_{lv}. The allocated bandwidth from network n to MT m via BS/AP s is denoted by b_{nms}. Let $B = [b_{nms}]$ be a matrix of allocated bandwidth from network $n \in \mathcal{N}$ to MT $m \in \mathcal{M}$ through BS/AP $s \in \mathcal{S}_n$, where $b_{nms} = 0$ if MT $m \notin \mathcal{M}_{ns}$ and for single-network MT if $a_m \neq ns$. For subscribers of a given network, let M_{lvk} denote the number of existing calls of service type v and service class l in service area k and C_{lvk} is the maximum number of calls of each service type v and service class l which can be supported in each service area k for subscribers of the given network. It is assumed that a call admission control procedure is in place, which guarantees that $M_{lvk} \leq C_{lvk}$, such that feasible resource allocation solutions exist with sufficient resources for a target call traffic load.

The arrival process of both new and handoff calls of service type v and class l to service area k is modeled by a Poisson process with parameter υ_{lvk}. A two-stage hyper-exponential distribution is used to approximate the PDF of the video call duration, T_c^{lv}, with mean $\bar{T}_c^{lv}$, which is given by [60]

$$f_{T_c^{lv}}(t) = \frac{a_{lv}}{a_{lv}+1} \cdot \frac{a_{lv}}{\bar{T}_c^{lv}} \cdot e^{-\frac{a_{lv}}{\bar{T}_c^{lv}}t} + \frac{1}{a_{lv}+1} \cdot \frac{1}{a_{lv}\bar{T}_c^{lv}} \cdot e^{-\frac{1}{a_{lv}\bar{T}_c^{lv}}t}, \quad a_{lv} \geq 1, t \geq 0.$$

$$(4.1)$$

4.2.4 Mobility Models and Channel Holding Time

The user residence time within service area k is modeled by an exponential distribution with mean $\bar{T}_r^k$. Hence, the channel holding time for a given service type v with service class l in service area k, $T_h^{lvk} = \min(T_c^{lv}, T_r^k)$, has a PDF that is given by

$$
f_{T_h^{lvk}}(t) = \frac{a_{lv}}{a_{lv}+1} \cdot \left(\frac{1}{\bar{T}_r^k} + \frac{a_{lv}}{\bar{T}_c^{lv}} \right) \cdot e^{-\left(\frac{1}{\bar{T}_r^k} + \frac{a_{lv}}{\bar{T}_c^{lv}} \right)t}
$$
$$
+ \frac{1}{a_{lv}+1} \cdot \left(\frac{1}{\bar{T}_r^k} + \frac{1}{a_{lv}\bar{T}_c^{lv}} \right) \cdot e^{-\left(\frac{1}{\bar{T}_r^k} + \frac{1}{a_{lv}\bar{T}_c^{lv}} \right)t}, \qquad t \geq 0. \quad (4.2)
$$

4.3 Centralized Optimal Resource Allocation (CORA)

In this section, the radio resource allocation problem is formulated for MTs with single-network and multi-homing services in the heterogeneous wireless access medium. Based on the problem formulation, a centralized optimal resource allocation (CORA) algorithm is then presented.

4.3.1 Problem Formulation

The utility of network n allocating bandwidth b_{nms} to MT m via BS/AP s, $u_{nms}(b_{nms})$, is given by
$$
u_{nms}(b_{nms}) = \ln(1 + \eta_1 b_{nms}) - \eta_2(1 - p_{nms})b_{nms} \quad (4.3)
$$

where η_1 and η_2 are used for scalability of b_{nms} [57].

Given a network assignment vector A, the overall resource allocation objective of all networks in the geographical region is to determine the optimal bandwidth allocation $b_{nms}, \forall n \in \mathcal{N}, m \in \mathcal{M}_{ns}, s \in \mathcal{S}_n$, which maximizes the total utility in the region, U, given by

$$
U = \sum_{n=1}^{N} \sum_{s=1}^{S_n} \sum_{m \in \mathcal{M}_{ns}} u_{nms}(b_{nms}). \quad (4.4)
$$

The allocated resources from network n BS/AP s should satisfy the BS/AP capacity constraint given by

$$\sum_{m \in \mathcal{M}_{ns}} b_{nms} \leq C_n, \quad \forall s \in \mathcal{S}_n, n \in \mathcal{N}. \tag{4.5}$$

Given a network assignment vector A, for MTs with single-network service, the allocated resources from the assigned network n BS/AP $s \in \mathcal{S}_{nk}$ to MT $m \in \mathcal{M}_{1k}$ in service area k should satisfy the application required bandwidth, given by

$$B_m^{\min} \leq b_{nms} \leq B_m^{\max}, \quad \forall m \in \mathcal{M}_{1k}, k \in \mathcal{K}. \tag{4.6}$$

While for MTs with multi-homing service, the total allocated resources from all available BSs/APs in $\mathcal{S}_k$ to MT $m \in \mathcal{M}_{2k}$ in service area k should satisfy the application total required bandwidth, which is given by

$$B_m^{\min} \leq \sum_{n \in \mathcal{N}_k} \sum_{s \in \mathcal{S}_{nk}} b_{nms} \leq B_m^{\max}, \quad \forall m \in \mathcal{M}_{2k}, k \in \mathcal{K}. \tag{4.7}$$

In order to determine the optimal network assignment vector A and the corresponding optimal bandwidth allocation matrix B for single-network and multi-homing MTs, the radio resource allocation problem is expressed by the following optimization problem

$$\max_A \{\max_{B \geq 0} \quad U$$
$$s.t. \quad (4.5) - (4.7).\} \tag{4.8}$$

The radio resource allocation problem for a given network assignment vector (i.e. the inner maximization problem of (4.8)) is a convex optimization problem that can be solved efficiently using polynomial time algorithms [6]. However, finding the optimal vector A (i.e. the outer maximization problem of (4.8)) incurs high computational complexity. In a given service area k with a total of $|\mathcal{M}_{1k}|$ MTs with single-network service and $|\mathcal{S}_k|$ BSs/APs available from different networks, there exist $|\mathcal{S}_k|^{|\mathcal{M}_{1k}|}$ distinct assignment vectors. As a result, the total number of distinct assignment vectors in the whole geographical region is $\prod_k |\mathcal{S}_k|^{|\mathcal{M}_{1k}|}$. For instance, consider one service area with a total of 50 MTs with single-network service and 3 BSs/APs having overlapped coverage. A total of $3^{50} = 7 * 10^{23}$ distinct network assignments exist in this service area. For the whole geographical region, it is expected that the inner maximization problem of (4.8) needs to be solved for a huge number of times so as to determine the optimal radio resource allocation (i.e. the optimal network assignment vector A and bandwidth allocation matrix B). As a result, it is desirable to develop a less complex formulation rather than the max-max formulation of problem (4.8). Towards this end, a binary assignment variable x_{nms} is introduced [59], that is determined from the network assignment vector A for MT $m \in \mathcal{M}_{1k}$ by

$$x_{nms} = \begin{cases} 1, & \text{if } a_m = ns \\ 0, & \text{otherwise.} \end{cases} \qquad (4.9)$$

while $x_{nms} = 1$ for MTs with multi-homing service in service area k for all $s \in S_k$. Using the binary assignment variable, the problem of (4.8) can be reformulated as

$$\max_{x_{nms}, b_{nms} \geq 0} \sum_{n=1}^{N} \sum_{s=1}^{S_n} \sum_{m \in \mathcal{M}_{ns}} \{\ln(1 + \eta_1 x_{nms} b_{nms}) - \eta_2(1 - p_{nms}) x_{nms} b_{nms}\}$$

$$s.t. \quad \sum_{m \in \mathcal{M}_{ns}} x_{nms} b_{nms} \leq C_n, \qquad \forall s \in \mathcal{S}_n, n \in \mathcal{N}$$

$$B_m^{\min} \leq \sum_{n=1}^{N} \sum_{s \in \mathcal{S}_{nk}} x_{nms} b_{nms} \leq B_m^{\max}, \qquad \forall m \in \mathcal{M}_k, k \in \mathcal{K} \quad (4.10)$$

$$x_{nms} \in \{0, 1\}, \qquad \forall m \in \mathcal{M}_{1k}, n \in \mathcal{N}_k, s \in \mathcal{S}_{nk}, k \in \mathcal{K}$$

$$\sum_{n=1}^{N} \sum_{s \in \mathcal{S}_{nk}} x_{nms} = 1, \qquad \forall m \in \mathcal{M}_{1k}, k \in \mathcal{K}$$

$$x_{nms} = 1, \qquad \forall m \in \mathcal{M}_{2k}, n \in \mathcal{N}_k, s \in \mathcal{S}_{nk}, k \in \mathcal{K}.$$

The fourth constraint ensures that an MT with single-network service is assigned to one and only one BS/AP available at its location, while the last constraint allows an MT with multi-homing service to obtain its required bandwidth from all wireless networks available at its location. The problem of (4.10) is a non-convex mixed integer non-linear programming (MINLP) problem. In general, MINLP problems combine the difficulty of optimizing over integer variables with the handling of non-linear functions which makes them difficult to solve [9]. This is especially true when the objective and/or constraint functions are non-convex, which is the case in (4.10). Several new methods are proposed recently for solving MINLP problems [22]. Two classes of algorithms that solve MINLP problems can be distinguished. The first class includes deterministic algorithms such as branch and bound, outer approximation, generalized benders decomposition, and extended cutting plane [9, 22]. Non-convexities in MINLP problems can be addressed by global optimization approaches which are developed using convex envelopes or under-estimators to formulate lower-bounding convex MINLP problems [22]. One example of deterministic global optimization methods for MINLP problems is branch and reduce [65], and other methods can be found in [22]. The second class of MINLP algorithms includes stochastic (heuristic) optimization algorithms such as the extended ant colony optimization [54].

The different algorithms of solving MINLP problems have been available through many solvers [10]. Deterministic solvers that claim to guarantee global optimality for non-convex general MINLP problems include AlphaBB, BARON, COUENNE, and LINDOGLOBAL [10]. On the other hand, stochastic solvers include MIDACO

[55], however there is no guarantee for global optimality [10]. The BARON solver
[53], which is available through GAMS [1], has proven to be the most robust one
among the currently available global solvers [44]. The BARON solver implements
deterministic global optimization algorithms which integrate conventional branch
and bound with a wide variety of range reduction tests [53]. The BARON solver
guarantees to provide global optima under fairly general assumptions which include
the availability of finite lower and upper bounds on the variables and their expressions
in the MINLP to be solved [53]. Hence, to solve the radio resource allocation problem
(4.10), we use the BARON solver through GAMS. The GDXMRW utilities [19] are
used to create an interface between GAMS and MATLAB in order to make use of
GAMS as a powerful optimization platform and the MATLAB visualization tools.

Figure 4.2 illustrates a centralized implementation of the radio resource allocation
(CORA) algorithm based on the formulation of (4.10). In the CORA algorithm, each
MT reports to all BSs/APs available at its location about its service type, service
class, and home network using its multiple radio interfaces. This information then
is made available to the central resource manager via different BSs/APs. Hence, the
central resource manager has the information regarding the service area k for each
MT, MT minimum and maximum required bandwidth, and MT priority parameter.
Given the transmission capacities of all the BSs/APs, the central resource manager
solves (4.10) so as to determine the optimal network assignment and bandwidth
allocations for new incoming MTs with single-network and multi-homing services,
updates bandwidth allocations, and initiates vertical handovers for existing MTs if
necessary.

4.3.2 Numerical Results and Discussion

This section presents numerical results for problem (4.10) using the BARON/GAMS
solver. Consider the simplified system model given in Fig. 4.3. We study the radio
resource allocation in service area 2 which is covered by the WiMAX (network 1)
and cellular network (network 2). For the service area under consideration, let the
transmission capacity of each network BS be 4 Mbps for network 1 and 1.248 Mbps
for network 2. The transmission capacities of different BSs are chosen such that they
can support a total of 12 MTs with VBR calls of required bandwidth $B_m \in [64, 128]$
kbps of single-network service, and a total of 17 MTs with VBR calls of required
bandwidth $B_m \in [256, 512]$ Kbps of multi-homing service. The number of subscribers from network n with service v is given by M_{nv}, where $v = 1$ represents
a single-network service while $v = 2$ represents a multi-homing service. With
$M_{11} = 6$, $M_{21} = 6$, $M_{22} = 8$, we vary the number of network 1 subscribers with
multi-homing service, M_{12}, in order to study the performance of the CORA algorithm as the call traffic load of the subscribers of the network with the larger capacity
varies. Using the priority parameter p_{nms}, the two networks set different costs on
their resources. Since the cellular network (network 2) has a smaller transmission
capacity than the WiMAX (network 1), it sets a higher cost on its resources so that it

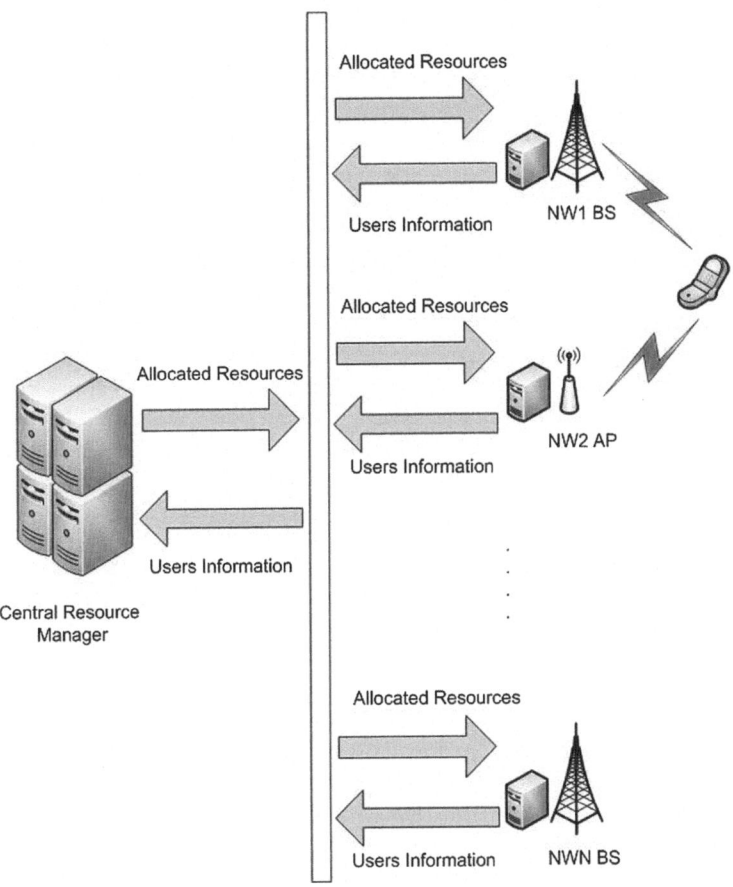

Fig. 4.2   Centralized implementation of the CORA algorithm

can devote its resources to its own subscribers [28]. As a result, let $p_{1m1} = 0.8$ and $p_{2m1} = 0.6$ for network users, while $p_{nm1} = 1$ for network subscribers with $n \in \mathcal{N}$. Let η_1 and η_2 equal 1. Let the number of assigned subscribers of network n, with single-network service, to network n' be $L_{nn'}$.

Figure 4.4 shows the allocated resources per call for MTs with single-network service versus the number M_{12} of network 1 subscribers with multi-homing service. As M_{12} increases, the allocated bandwidth for network 2 subscribers is reduced first towards the minimum required bandwidth. This is because network 2 subscribers rely heavily on network 1 resources in addition to their home network in order to support

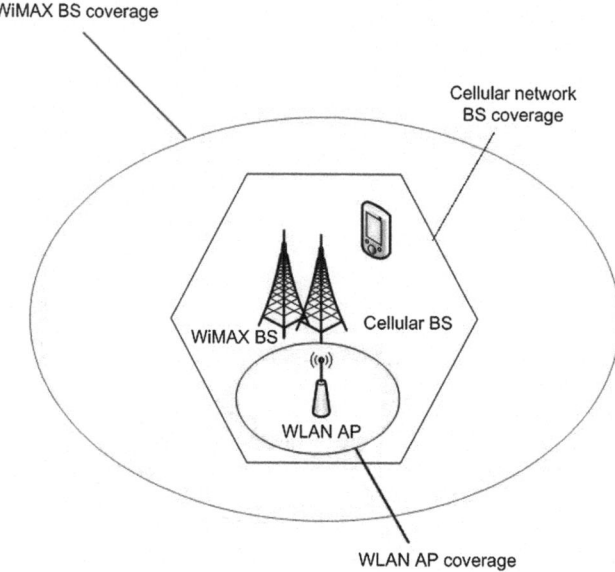

Fig. 4.3 Service areas under consideration

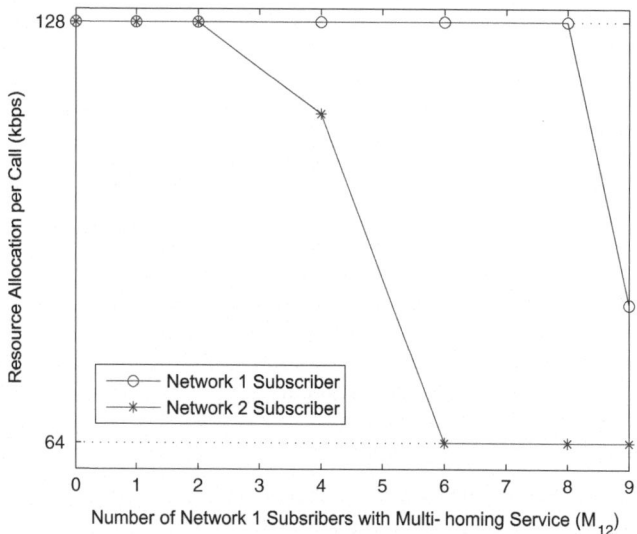

Fig. 4.4 Radio resource allocation for MTs with single-network service

their high required bandwidth, while network 1 gives a higher priority to its own subscribers on its resources using the priority mechanism. The allocated resources to network 1 subscribers is then reduced so as to accommodate more multi-homing

Table 4.1 Network assignments for network 1 and network 2 subscribers with single-network service

M_{12}	L_{11}	L_{12}	L_{21}	L_{22}
0	6	0	6	0
1	6	0	6	0
2	6	0	6	0
4	6	0	4	2
6	6	0	3	3
8	6	0	3	3
9	6	0	2	4

subscribers (M_{12}) from this network. Overall, the resource allocation guarantees the desired bandwidth range for the VBR calls.

Table 4.1 shows the numbers of MTs with single-network services assigned to each BS/AP for network 1 and network 2 subscribers versus the number M_{12} of network 1 subscribers with multi-homing service. Due to the larger capacity of network 1, its subscribers are always assigned to their home network (L_{11}) which provides them with high allocated bandwidth (refer to Fig. 4.4). As for network 2 subscribers, their network assignment varies with M_{12}. At a small number of M_{12} (from 0 to 2), all network 2 subscribers with single-network service are assigned to network 1 (L_{21}), as it provides them with their maximum required bandwidth (refer to Fig. 4.4). As the call traffic load increases in network 1 (due to an increase in M_{12}), more subscribers from network 2 are assigned to their home network (L_{22}), as network 1 gives higher priority to its own subscribers on its resources.

Figure 4.5 shows the allocated resources per call for MTs with multi-homing service from each available network versus the number M_{12} of network 1 subscribers with multi-homing service. The total allocated bandwidth to network 1 subscribers ($N1$) comes from network 1 ($N1 - 1$). The allocated bandwidth from network 2 ($N2 - 1$) is zero, since network 2 devotes its resources to support its own subscribers using the priority parameter p_{2m1}. The total allocated resources per call for network 1 subscribers ($N1$) decreases with M_{12} towards the minimum required bandwidth to accommodate more subscribers. For network 2 subscribers, the allocated resources from network 1 ($N1 - 2$) decreases as M_{12} increases, since network 1 uses its resources to support its own subscribers. This is compensated by an increase in the resource allocation from network 2 ($N2 - 2$) to improve the allocated resources to its own subscribers. However, for $M_{12} > 2$, network 2 decreases its allocated resources to its subscribers with multi-homing service, since more single-network subscribers are assigned to its BS (refer to Table 4.1). As a result, the total allocated resources per call for network 2 subscribers ($N2$) decreases with M_{12} towards the minimum required bandwidth. The total allocated resources per call for network 1 and network 2 subscribers with multi-homing services ($N1$ and $N2$ respectively) are within the desired bandwidth range for the VBR calls.

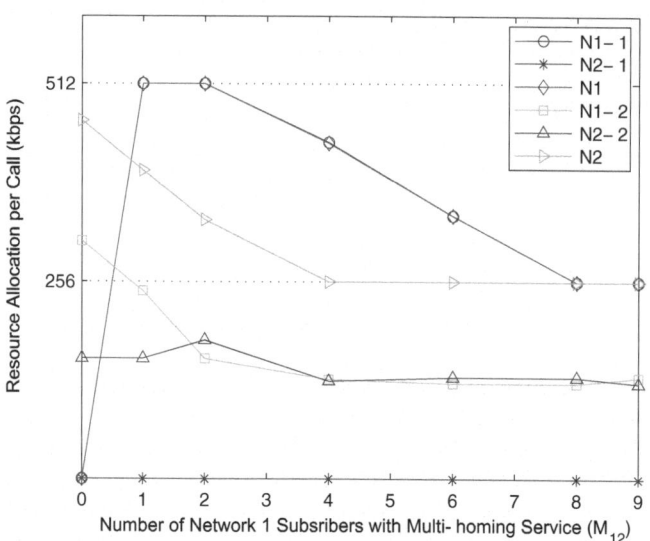

Fig. 4.5 Radio resource allocation for MTs with multi-homing service

4.4 Decentralized Sub-Optimal Resource Allocation (DSRA)

In this section, a decentralized sub-optimal resource allocation (DSRA) algorithm is presented for the radio resource allocation problem. The DSRA algorithm is desirable when different networks are operated by different service providers.

In problem (4.8), if the network assignment vector A is known, the problem is reduced to finding the optimal bandwidth allocation matrix B which is a convex optimization problem that can be solved in a decentralized manner using the decomposition approach discussed in Chap. 2. In order to find the network assignment vector A, call traffic load prediction and network cooperation concepts presented in Chap. 3 can be employed. Hence, in the DSRA algorithm, time is partitioned into a set of periods $\mathcal{T} = \{T_1, T_2, \ldots, T_j, \ldots\}$ of constant duration τ. At each BS/AP, the call traffic load at current period, T_j, is used to predict the call traffic load during the next period, T_{j+1}. By exchanging their predicted call traffic load information for the next period, cooperative BSs/APs can determine the distribution of the total call traffic load in the geographical region (i.e. network assignment vector A) for the next period, T_{j+1}. Based on the predicted call traffic load, every BS/AP broadcasts a parameter (a predicted link access price) which enables incoming and existing MTs to perform network selection and bandwidth request without the need for a central resource manager. As in Chap. 3, the traffic load prediction is a probabilistic one which ensures that the prediction error is lower than a target value ϵ, and ϵ is chosen based on the target call blocking probability of the system. The DSRA algorithm can be carried out in the following 8 steps.

Step 1: For clarity of presentation, we focus our discussion in steps 1–3 on one network subscribers, and the same steps hold for subscribers of other networks.

Consider video calls of service type v and class l in service area k. Let T_{lvk}^{j} be a time vector of call arrival events for calls of service type v and service class l in service area k during period T_j. With a call arrival event at time instant $t_a^{j} \in T_{lvk}^{j}$, $a = \{1, 2, \ldots, |T_{lvk}^{j}|\}$, in period T_j, the number of calls at the time instant, $M_{lvk}(t_a^{j})$, is used by the BSs/APs in the service area to probabilistically predict the number of calls at time instant $t_a^{j} + \tau$ in the next time period T_{j+1}. The predicted number is given by $\widetilde{M}_{lvk}(t_a^{j} + \tau)$. As the number of calls at t, $M_{lvk}(t)$, is a random variable, using the probability distribution of $M_{lvk}(t_a^{j} + \tau)$ given $M_{lvk}(t_a^{j})$, we can represent $\widetilde{M}_{lvk}(t_a^{j} + \tau)$ by a design parameter ϵ_{lvk}, such that

$$Pr[M_{lvk}(t_a^{j} + \tau) > \widetilde{M}_{lvk}(t_a^{j} + \tau)|M_{lvk}(t_a^{j})] \leq \epsilon_{lvk}, \quad \forall v, l \in \mathcal{L}, k \in \mathcal{K}. \quad (4.11)$$

Similar to ϵ_{lk}^{n} in Chap. 3, the design parameter $\epsilon_{lvk} \in [0, 1]$ denotes the probability that $M_{lvk}(t_a^{j} + \tau)$ exceeds the predicted number $\widetilde{M}_{lvk}(t_a^{j} + \tau)$. The predicted number $\widetilde{M}_{lvk}(t_a^{j} + \tau)$ can be determined using the conditional probability mass function (PMF) of $M_{lvk}(t_a^{j} + \tau)$ given $M_{lvk}(t_a^{j})$, $P_{M_{lvk}(t_a^{j}+\tau)|M_{lvk}(t_a^{j})}(i)$. Again, the transient distribution of the $M/G/\infty$ model [40] can be used to calculate $P_{M_{lvk}(t_a^{j}+\tau)|M_{lvk}(t_a^{j})}(i)$, since call arrivals follow a Poisson process, the channel holding time follows a general distribution, and all calls are served simultaneously without queuing. We redefine p_τ^{lk} and q_τ^{lk} introduced in Chap. 3, under the assumption of stationary call arrival and departure processes, to be:

- p_τ^{lvk}—The probability that a call of service class l and service type v which is in service area k at time t_a^{j} is still present in the same service area at time $t_a^{j} + \tau$;
- q_τ^{lvk}—The probability that a call of service class l and service type v that arrives in service area k during $(t_a^{j}, t_a^{j} + \tau]$ is still present at the same service area at time $t_a^{j} + \tau$;

while $X_B(\kappa, \alpha)$ and $X_P(\alpha)$ are the same as in Chap. 3,

- $X_B(\kappa, \alpha)$—A binomial random variable with parameters κ and α;
- $X_P(\alpha)$—A Poisson random variable with mean α.

At time instant t_a^{j}, given the number of calls, $M_{lvk}(t_a^{j})$, we have [40]

$$M_{lvk}(t_a^{j} + \tau) =_d X_B(M_{lvk}(t_a^{j}), p_\tau^{lvk}) + X_P(v_{lvk}^{n}\tau q_\tau^{lvk}) \quad (4.12)$$

where v_{lvk}^{n} denotes the arrival rate of new and handoff calls to network n in service area k. In order to determine v_{lvk}^{n} for BS/AP of network n, a BS/AP can count the number of its new call arrivals to service area k (excluding vertical handoff calls, since these calls are not arrivals to service area k) and divide it by the total elapsed time. In (4.12), the probabilities p_τ^{lvk} and q_τ^{lvk} are given by [40]

$$p_\tau^{lvk} = \frac{1}{E[T_h^{lvk}]} \int_\tau^\infty (1 - F_{T_h^{lvk}}(s))ds. \tag{4.13}$$

$$q_\tau^{lvk} = \frac{E[T_h^{lvk}]}{\tau}(1 - p_\tau^{lvk}) \tag{4.14}$$

where $E[T_h^{lvk}]$ denotes the average channel holding time which can be calculated using (4.2). From (4.12), $P_{M_{lvk}(t_a^j+\tau)|M_{lvk}(t_a^j)}(i)$ can be found, and hence $\widetilde{M}_{lvk}(t_a^j+\tau)$ can be calculated using (4.11) as the minimum integer which satisfies

$$\sum_{i=0}^{\widetilde{M}_{lvk}(t_a^j+\tau)} P_{M_{lvk}(t_a^j+\tau)|M_{lvk}(t_a^j)}(i) \geq (1 - \epsilon_{lvk}), \quad \forall v, l \in \mathcal{L}, k \in \mathcal{K}. \tag{4.15}$$

Step 2: Each BS/AP in service area k records the predicted values of $\widetilde{M}_{lvk}(t_a^j+\tau)$, $\forall v, l \in \mathcal{L}, k \in \mathcal{K}$ and $a = \{1, 2, \ldots, |\mathcal{T}_{lvk}^j|\}$ in a vector $\mathcal{M}_{lvk}^{j+1}$.

Step 3: At the beginning of period T_{j+1}, the maximum predicted number of calls of each service type v and service class l in each service area k during T_{j+1}, $\widetilde{M}_{lvk}(T_{j+1})$, can be found using $\mathcal{M}_{lvk}^{j+1}$. That is, $\widetilde{M}_{lvk}(T_{j+1}) = \max(\mathcal{M}_{lvk}^{j+1})$ if it is less than or equal to C_{lvk}, otherwise $\widetilde{M}_{lvk}(T_{j+1}) = C_{lvk}$. This guarantees that for $\widetilde{M}_{lvk}(T_{j+1}) \leq C_{lvk}$ we have

$$Pr[M_{lvk}(t_a^{j+1}) > \widetilde{M}_{lvk}(T_{j+1})] \leq \epsilon_{lvk},$$
$$\forall v, l \in \mathcal{L}, k \in \mathcal{K}, a \in \{1, 2, \ldots, |\mathcal{T}_{lvk}^{j+1}|\}. \tag{4.16}$$

Step 4: The cooperating BSs/APs in the geographical region exchange their information regarding $\widetilde{M}_{lvk}(T_{j+1})$ $\forall v, l \in \mathcal{L}, k \in \mathcal{K}$ for all subscribers. As a result, $\mathcal{M}_{lvk}$ can be determined and hence problem (4.10) can be solved at each BS/AP so as to determine the binary assignment variable x_{nms}^{j+1} for all MTs with single-network service in the geographical region during T_{j+1} and the corresponding bandwidth allocation matrix B^{j+1}. Therefore, the network assignment vector A^{j+1} for single-network MTs during T_{j+1} can be determined. Based on the network assignment vector A^{j+1}, each BS/AP s can determine the maximum number of MTs with single-network calls and service class l in service area k which can be supported by this BS/AP during T_{j+1}, f_{lks}^{j+1}, $\forall l \in \mathcal{L}, k \in \mathcal{K}$, given B^{j+1}.

Step 5: Given the network assignment vector A^{j+1}, during T_{j+1}, calculated in step 4, problem (4.8) is reduced to

$$\max_{B \geq 0} U$$
$$s.t. \quad (4.5) - (4.7). \tag{4.17}$$

Problem (4.17) is a convex optimization problem, on which full dual decomposition can be applied (this helps in the decentralized resource allocation as described in

the next step). As in Chap. 2, in order to apply full dual decomposition, we first find
the Lagrangian function, $L(B, \lambda, \nu^{(1)}, \nu^{(2)}, \mu^{(1)}, \mu^{(2)})$, of (4.17), where $\lambda = (\lambda_{ns} : n \in \mathcal{N}, s \in \mathcal{S}_n)$ is defined to be a matrix of Lagrange multipliers corresponding to
capacity constraint (4.5), and $\lambda_{ns} \geq 0, \nu^{(1)} = (\nu_m^{(1)} : m \in \mathcal{M}_{1k}, \forall k \in \mathcal{K})$ and $\nu^{(2)} = (\nu_m^{(2)} : m \in \mathcal{M}_{1k}, \forall k \in \mathcal{K})$ are vectors of Lagrangian multipliers corresponding to
the maximum and minimum required bandwidth constraints of (4.6) for MTs with
single-network service and $\nu_m^{(1)}, \nu_m^{(2)} \geq 0$, and $\mu^{(1)} = (\mu_m^{(1)} : m \in \mathcal{M}_{2k}, \forall k \in \mathcal{K})$,
$\mu^{(2)} = (\mu_m^{(2)} : m \in \mathcal{M}_{2k}, \forall k \in \mathcal{K})$ are vectors of lagrange multipliers corresponding
to the required bandwidth constraints for MTs with multi-homing service (4.7) and
$\mu_m^{(1)}, \mu_m^{(2)} \geq 0$. The dual function then is given by

$$h(\lambda, \nu^{(1)}, \nu^{(2)}, \mu^{(1)}, \mu^{(2)}) = \max_{B \geq 0} L(B, \lambda, \nu^{(1)}, \nu^{(2)}, \mu^{(1)}, \mu^{(2)}) \qquad (4.18)$$

and the dual problem corresponding to the primal problem of (4.17) is given by

$$\min_{(\lambda, \nu^{(1)}, \nu^{(2)}, \mu^{(1)}, \mu^{(2)}) \geq 0} h(\lambda, \nu^{(1)}, \nu^{(2)}, \mu^{(1)}, \mu^{(2)}). \qquad (4.19)$$

The maximization problem (4.18) gives the bandwidth allocation matrix B for fixed
value of the Lagrangian multipliers, which can be solved using the KKT conditions,
and hence we have

$$b_{nms} = \left[\left(\frac{\eta_1}{\lambda_{ns} + (\nu_m^{(1)} - \nu_m^{(2)}) + \eta_2(1 - p_{nms})} - 1 \right) / \eta_1 \right]^+, \quad \forall m \in \bigcup_k \mathcal{M}_{1k} \qquad (4.20)$$

$$b_{nms} = \left[\left(\frac{\eta_1}{\lambda_{ns} + (\mu_m^{(1)} - \mu_m^{(2)}) + \eta_2(1 - p_{nms})} - 1 \right) / \eta_1 \right]^+, \quad \forall m \in \bigcup_k \mathcal{M}_{2k} \qquad (4.21)$$

The optimal values of the Lagrangian multipliers which result in the optimal bandwidth allocation can be found by solving the dual problem of (4.19). For a differentiable dual function, a gradient descent method can be applied to determine the
optimum values for the Lagrangian multipliers, which results in

$$\lambda_{ns}(i+1) = \left[\lambda_{ns}(i) - \alpha_1 \left(C_n - \sum_{m \in \mathcal{M}_{ns}} b_{nms}(i) \right) \right]^+ \qquad (4.22)$$

$$\nu_m^{(1)}(i+1) = \left[\nu_m^{(1)}(i) - \alpha_2 (B_m^{\max} - b_{nms}(i)) \right]^+ \qquad (4.23)$$

$$\nu_m^{(2)}(i+1) = \left[\nu_m^{(2)}(i) - \alpha_3 (b_{nms}(i) - B_m^{\min}) \right]^+ \qquad (4.24)$$

$$\mu_m^{(1)}(i+1) = \left[\mu_m^{(1)}(i) - \alpha_4 \left(B_m^{\max} - \sum_{n=1}^{N} \sum_{s=1}^{S_n} b_{nms}(i) \right) \right]^+ \qquad (4.25)$$

$$\mu_m^{(2)}(i+1) = \left[\mu_m^{(2)}(i) - \alpha_5 \left(\sum_{n=1}^{N} \sum_{s=1}^{S_n} b_{nms}(i) - B_m^{\min} \right) \right]^+ \qquad (4.26)$$

where i is the iteration index and α_j with $j = \{1, \ldots, 5\}$ is a fixed sufficiently small step size. Convergence towards the optimal solution is guaranteed as the gradient of (4.19) satisfies the Lipchitz continuity condition.

As in Chaps. 2 and 3, λ_{ns} is a link access price that is used as an indication of the capacity limitation experienced by each network BS/AP, while $\mu_m^{(1)}$ and $\mu_m^{(2)}$ are used by MTs with multi-homing calls to guarantee that the total bandwidth allocated from all BSs/APs satisfy the call total required bandwidth. On the other hand, $\nu_m^{(1)}$ and $\nu_m^{(2)}$ are used by MTs with single-network calls to guarantee that the bandwidth allocated from the assigned network satisfies the call required bandwidth.

Given the predicted maximum number of calls during T_{j+1}, $\widetilde{M}_{lvk}(T_{j+1})$ $\forall v, l \in \mathcal{L}, k \in \mathcal{K}, n \in \mathcal{N}$, each BS/AP can determine its predicted link access price value $\widetilde{\lambda}_{ns}^{j+1}$ using the BARON solver while solving (4.10) at the beginning of T_{j+1} using $\widetilde{M}_{lvk}(T_{j+1})$.

Step 6: At the beginning of T_{j+1}, each BS/AP updates its link access price value with $\widetilde{\lambda}_{ns}^{j+1}$ and this value is fixed over T_{j+1}, independent of call arrivals to and departures from different service areas, and is broadcasted on the BS/AP ID beacon. In addition, a flag bit, fb_{lks}, is set to 1 if $M_{lvk} < f_{lks}$ and is broadcasted by each BS/AP s on its ID beacon to denote that a new incoming call from subscribers of a given network with single-network service and service class l in service area k can be admitted by the BS/AP. Otherwise, $fb_{lks} = 0$.

The fixed link access price values, $\widetilde{\lambda}_{ns}^{j+1}$ $\forall n \in \mathcal{N}, s \in \mathcal{S}_n$ which are broadcasted during T_{j+1}, distribute the radio resources of all networks exactly over the maximum predicted number of calls $\widetilde{M}_{lvk}(T_{j+1})$ $\forall v, l \in \mathcal{L}, k \in \mathcal{K}$. Hence, during T_{j+1}, when $M_{lvk} = \widetilde{M}_{lvk}(T_{j+1})$, any incoming call from subscribers of a given network with service type v and service class l in service area k will be blocked. Hence, similar to ϵ_{lk}^n, from (4.11), ϵ_{lvk} is the upper bound of the call blocking probability for subscribers of a given network, given that $\widetilde{M}_{lvk}(T_{j+1}) \leq C_{lvk}$. Otherwise, $\widetilde{M}_{lvk}(T_{j+1}) = C_{lvk}$, and both the CORA and DSRA algorithms achieve the same call blocking probability.

Step 7: An incoming MT to service area k during T_{j+1} listens to the link access price values $\widetilde{\lambda}_{ns}^{j+1}$ $\forall n \in \mathcal{N}, s \in \mathcal{S}_n$ using its multiple radio interfaces. Based on its service type, the MT then performs the following.

First, consider MTs with single-network service. An MT, $m \in \mathcal{M}_{1k}$, uses the link access price values to solve for the allocated bandwidth from each BS/AP available at its location with $fb_{lks} = 1$. This can be done at MT, m, with a call from service class l in service area k, using the algorithm in Table 4.2, where I denotes the number of iterations required for the algorithm to converge to the required bandwidth allocation. Then, the MT orders the available BSs/APs based on the calculated bandwidth allocation from maximum to minimum. The MT asks the BS/AP with the maximum calculated bandwidth allocation for the b_{nms} resource allocation. The BS/AP provides the required bandwidth allocation if it has sufficient resources. Otherwise, the

Table 4.2 Calculation of bandwidth allocation from each available network BS/AP at MT m with single-network service

1: **Input:** $\widetilde{\lambda}_{ns}^{j+1} \ \forall n \in \mathcal{N}_k, s \in \mathcal{S}_{nk}, B_m, m \in \mathcal{M}$;
2: **Initialization:** $\nu_m^{(1)}(1) \geq 0; \nu_m^{(2)}(1) \geq 0$;
3: **for** $n \in \mathcal{N}_k$ **do**
4:     **for** $s \in \mathcal{S}_{nk}$ **do**
5:         **for** $i = 1 : I$ **do**
6:             $b_{nms}(i) = [(\frac{\eta_1}{\widetilde{\lambda}_{ns}^{j+1} + (\nu_m^{(1)}(i) - \nu_m^{(2)}(i)) + \eta_2(1 - p_{nms})} - 1)/\eta_1]^+$;
7:             $\nu_m^{(1)}(i+1) = [\nu_m^{(1)}(i) - \alpha_1(B_m^{\max} - b_{nms}(i))]^+$;
8:             $\nu_m^{(2)}(i+1) = [\nu_m^{(2)}(i) - \alpha_2(b_{nms}(i) - B_m^{\min})]^+$;
9:         **end for**
10:    **end for**
11: **end for**
12: **Output:** $b_{nms} \ \forall n \in \mathcal{N}_k, s \in \mathcal{S}_{nk}$.

Table 4.3 Calculation of bandwidth share from each available network BS/AP at MT m with multi-homing service

1: **Input:** $\widetilde{\lambda}_{ns}^{j+1} \ \forall n \in \mathcal{N}_k, s \in \mathcal{S}_{nk}, B_m, m \in \mathcal{M}$;
2: **Initialization:** $\mu_m^{(1)}(1) \geq 0; \mu_m^{(2)}(1) \geq 0$;
3: **for** $i = 1 : I$ **do**
4:     **for** $n \in \mathcal{N}_k$ **do**
5:         **for** $s \in \mathcal{S}_{nk}$ **do**
6:             $b_{nms}(i) = [(\frac{\eta_1}{\widetilde{\lambda}_{ns}^{j+1} + (\mu_m^{(1)}(i) - \mu_m^{(2)}(i)) + \eta_2(1 - p_{nms})} - 1)/\eta_1]^+$;
7:         **end for**
8:     **end for**
9:     $\mu_m^{(1)}(i+1) = [\mu_m^{(1)}(i) - \alpha_3(B_m^{\max} - \sum_{n=1}^{N}\sum_{s=1}^{S_n} b_{nms}(i))]^+$;
10:    $\mu_m^{(2)}(i+1) = [\mu_m^{(2)}(i) - \alpha_4(\sum_{n=1}^{N}\sum_{s=1}^{S_n} b_{nms}(i) - B_m^{\min})]^+$;
11: **end for**
12: **Output:** The required $b_{nms} \ \forall n \in \mathcal{N}_k, s \in \mathcal{S}_{nk}$.

incoming call is blocked. For MTs which are already in service, the link access price values $\widetilde{\lambda}_{ns}^{j+1} \ \forall n \in \mathcal{N}_k, s \in \mathcal{S}_{nk}$ with $fb_{lks} = 1$, are used at the beginning of T_{j+1} in a similar way as described before in order to perform a vertical handover if necessary.

Next, consider MTs with multi-homing services. During T_{j+1}, each MT in the geographical region, including both incoming and existing ones, uses the broadcasted link access price values received at its location to determine the required bandwidth share from each available BS/AP, such that the total amount of allocated resources from all the BSs/APs satisfies its required bandwidth. This is performed at MT, m, with service class l in service area k using the algorithm in Table 4.3. The MT then asks for the required bandwidth share b_{nms} from BS/AP s of network $n \ \forall n \in \mathcal{N}_k, s \in \mathcal{S}_{nk}$, which allocates the required bandwidth if it has sufficient resources. The incoming call is blocked if the total allocated resources from all BSs/APs do not satisfy its required bandwidth.

Step 8: Each MT reports to its serving BSs/APs its home network, service type, service class, and a list of the BS/AP IDs that the MT can receive signal from. This information is used by BSs/APs to predict $\widetilde{M}_{lvk}(T_{j+2}) \; \forall v, l \in \mathcal{L}, k \in \mathcal{K}$ for every network subscribers, during the next period T_{j+2} in order to update their link access price values at the beginning of T_{j+2}.

As in the PBRA algorithm, the link access price value for BSs/APs of different networks are updated every τ which should reflect some change in the call traffic load in the geographical region. Let δ_{lvk} be the minimum of durations to the arrival of a new call and to the departure of an existing call for service class l with service type v in service area k for subscribers of a given network. Define $\delta = \min(\delta_{lvk})$ $\forall l, v, k$ and subscribers of different networks. Thus, as a guideline, the time duration τ is chosen such that the probability $Pr[\delta < \tau]$ is less than a small threshold γ.

4.5 Simulation Results and Discussion

This section presents simulation results for the radio resource allocation problem in a heterogeneous wireless access medium for MTs with single-network and multi-homing services. Consider the geographical region given in Fig. 4.3. A single service class ($l = 1$) is considered for each service type v (single-network and multi-homing) and we study the performance of the proposed algorithms in the service area that is covered by the WiMAX and cellular network BSs ($k = 2$) in terms of the allocated resources per call and the call blocking probability. As a proof of concept, we only show the results of resource allocation for the cellular network subscribers. For simplicity, we consider a complete partitioning strategy for each network BS transmission capacity [42], where the total capacity of each BS is divided into two separate parts, dedicating to single-network and multi-homing services respectively.[1] The allocated transmission capacity from network n BS/AP to the service area under consideration for cellular network subscribers with service type v, C_{nv}, is given by $C_{11} = 1.344$, $C_{12} = 2.864$, $C_{21} = 0.576$, and $C_{22} = 2$ Mbps. The C_{nv} values can support a total of 30 VBR calls with required bandwidth allocation $B_m \in [0.064, 0.128]$ Mbps for single-network MTs, i.e. $C_{112} = 30$, and 19 VBR calls with required bandwidth allocation $B_m \in [0.256, 0.512]$ Mbps for multi-homing MTs, i.e. $C_{122} = 19$. The arrival process of new and handoff calls to the service area under consideration is modeled as a Poisson process with parameter v_{112} (call/min) for single-network MTs and v_{122} (call/min) for multi-homing MTs. The video call duration is modeled by a two-stage hyper-exponential distribution with the PDF given in (4.1) and $a_{1v} = 1$. The average call duration for single-network MTs $\bar{T}_c^{11}$ is 15 min and for multi-homing MTs $\bar{T}_c^{12}$ is 10 min. The user residence time in the service area under consideration follows an exponential distribution with an

---

[1] The numerical results in Sect. 4.3.2 investigates a complete sharing strategy for each BS/AP transmission capacity [42] where both service types can occupy up to the total capacity of each BS/AP.

average duration $\bar{T}_r = 20$ min [60]. The parameters η_1 and η_2 are both set to 1 [57]. The WiMAX and cellular networks set different costs on their resources using the priority parameter $p_{1m1} = 0.8$, $p_{2m1} = 0.6$ for network users, while $p_{nms} = 1$ for network subscribers [28]. The GDXMRW utilities [19] are used to create an interface between GAMS and MATLAB to make use of the BARON solver of GAMS in solving the optimization problem of (4.10) while using the MATLAB simulation and visualization tools.

4.5.1 Performance Comparison

In the following, the performance of the DSRA algorithm is compared to the CORA algorithm. While it is not appropriate for practical implementation when different networks are operated by different service providers, the CORA algorithm is used as a performance bound for the allocated resources per call and the call blocking probability. In the simulation, we set the upper bounds on call blocking probability ϵ_{112}, ϵ_{122} to 1 % and the prediction duration τ to 0.25, 0.5, and 1 min. We only show the results for single-network service and similar observations hold for multi-homing service.

Figure 4.6 shows performance comparison between the DSRA and CORA algorithms for MTs with single-network service versus the call arrival rate υ_{112}. Figure 4.6a shows the bandwidth allocation per call for MTs assigned to the WiMAX and MTs assigned to the cellular network. At a low call arrival rate, the predicted number of simultaneously present calls is low, which results in a high allocated bandwidth per call using the DSRA algorithm for different τ values. At a high call arrival rate, the predicted number of simultaneously present users is high, as a result less bandwidth is allocated to each call. Furthermore, less bandwidth is allocated per call for larger values of τ as explained in the next sub-section. Figure 4.6b shows that more MTs with single-network service are assigned to the WiMAX BS as compared to the cellular network BS due to the WiMAX BS larger capacity C_{11}. In Fig. 4.6c, using the CORA algorithm, there is no call blocking probability for $\upsilon_{112} < 1.6$ call/min. For call arrival rate $\upsilon_{112} < 2.2$ call/min, the DSRA algorithm does not exceed the target upper bound on call blocking probability of 1 %. For call arrival rate $\upsilon_{112} \geq 2.2$ call/min, the predicted number of calls simultaneously present in the service area under consideration is larger than C_{112}. Hence, according to the DSRA algorithm, the predicted number of calls is made equal to C_{112}, and both the DSRA and the CORA algorithms achieve the same call blocking probability.

4.5.2 Performance of the DSRA Algorithm

In the following, we study the performance of the DSRA algorithm versus its two design parameters, namely the upper bound on call blocking probability ϵ_{lvk} and the prediction duration τ. We only show the results for multi-homing service and the same observations hold for single-network service.

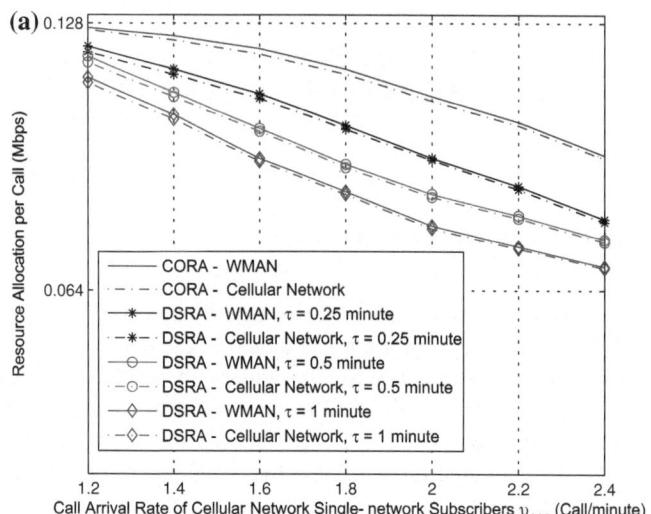

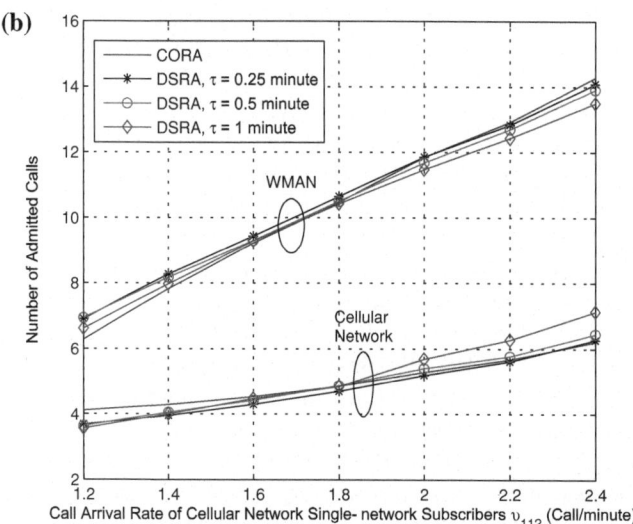

Fig. 4.6 Performance comparison for single-network service. **a** Resource allocation per call. **b** Number of admitted calls. **c** Call blocking probability. $\epsilon_{122} = 1\%$ and $\tau = 0.25, 0.5$, and 1 min

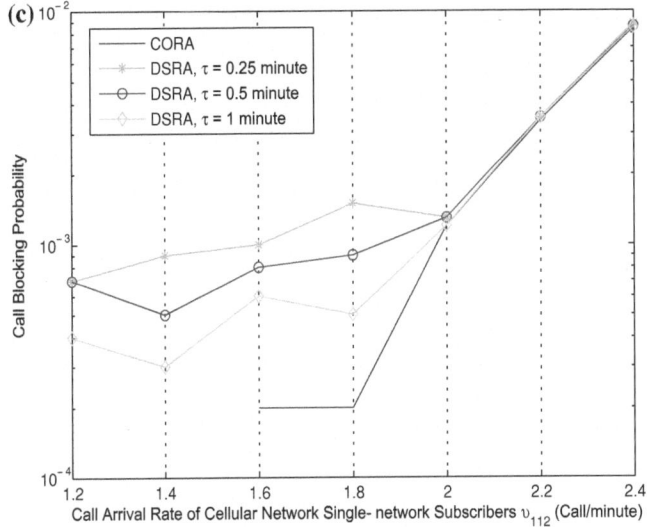

Fig. 4.6 (Continued)

Figure 4.7a plots the performance of the DSRA algorithm in terms of the amount of allocated resources per call and call blocking probability versus ϵ_{122}, with call arrival rate $\upsilon_{122} = 1.4$ call/min and $\tau = 1$ min. A small value of ϵ_{122} results in a low call blocking probability. However, this corresponds to a large number of predicted calls (and hence large BS/AP link access price values), which results in a small amount of resource allocation per call. On the other hand, a large value of ϵ_{122} results in a high call blocking probability and a large amount of resource allocation per call. Overall, the call blocking probability does not exceed its upper bound $\epsilon_{122} = 1$ %. The upper bound ϵ_{122} should be chosen to balance the trade-off between the allocated resources per call and the call blocking probability.

Figure 4.7b investigates the performance of the DSRA algorithm in terms of the amount of allocated resources per call and call blocking probability versus the prediction duration τ, with $\upsilon = 1.4$ call/min and $\epsilon_{122} = 1$ %. As τ increases, the DSRA algorithm updates the BS/AP link access price less frequently and hence a larger number of simultaneously present calls is predicted. As a result, the allocated resources per call is reduced. Also, simulation results indicate that the call blocking probability does not exceed its target upper bound $\epsilon_{122} = 1$ %.

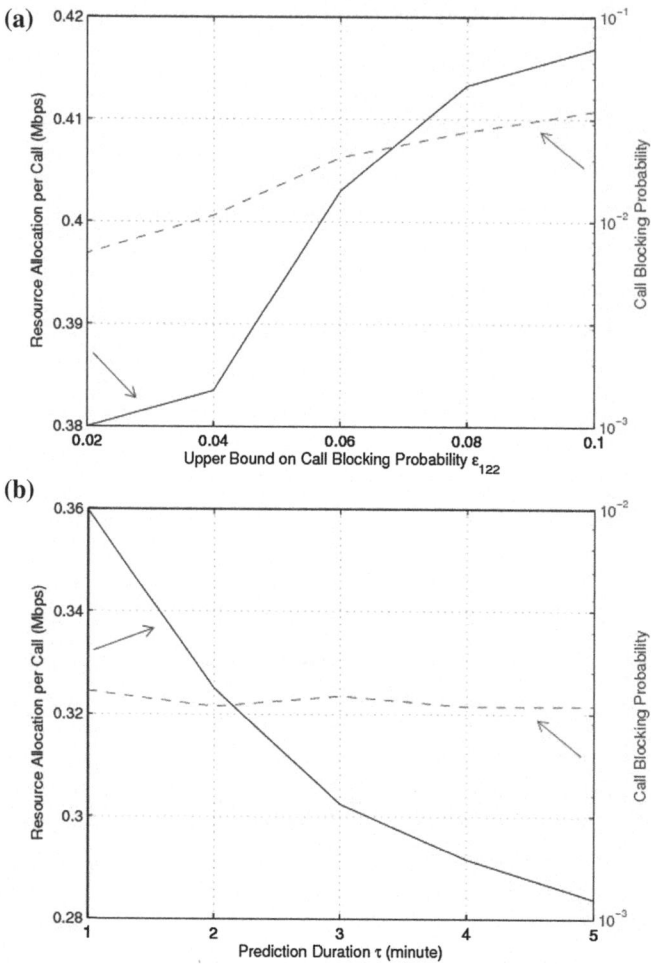

Fig. 4.7 The DSRA algorithm performance versus: **a** ϵ_{122}; **b** τ. $v_{122} = 1.4$ call/min and $\tau = 1$ min

4.6 Summary

In this chapter, a decentralized resource allocation algorithm is proposed for a heterogeneous wireless access medium to support MTs with single-network and multi-homing services. The algorithm gives MTs an active role in the resource allocation operation, such that an MT with single-network service can select the best wireless network available at its location and asks for its required bandwidth, while an MT with multi-homing service can determine the required bandwidth share from each network in order to satisfy its total required bandwidth. The resource allocation relies on concepts of short-term call traffic prediction and network cooperation in order to perform the decentralized resource allocation in an efficient manner. The algorithm has two design parameters, namely ϵ_{lvk} and τ, which should be properly chosen to strike a balance between the desired performance in terms of the allocated resources per call and the call blocking probability, and between the performance and implementation complexity.

Chapter 5
Conclusions and Future Directions

In this chapter, we summarize the main ideas and concepts presented in this brief and highlight future research directions.

5.1 Conclusions

In this brief, we have investigated radio resource allocation in heterogeneous wireless access medium. Based on the analysis and discussion provided throughout this brief, we present the following remarks.

- The heterogeneous wireless access medium creates various opportunities that can enhance perceived QoS for mobile users. However, it is necessary to develop new radio resource allocation mechanisms in order to satisfy required QoS of different calls while at the same time make efficient utilization of the available resources from different networks;
- One important aspect of radio resource allocation mechanisms is the need to operate in a decentralized manner (i.e. without a central resource manager). This adds a desirable flexibility to the radio resource allocation and avoids many complications associated with the centralized solutions (e.g., creating a single point of failure);
- The radio resource allocation mechanisms should give each network a higher priority in allocating its resources to its own subscribers as compared to other users. In this sense, network users can enjoy their maximum QoS but not at the expense of the network subscribers;
- Co-existence of single-network and multi-homing services in the heterogeneous wireless access medium should be considered. Hence, a radio resource allocation mechanism is to find the network assignment for MTs with single-network calls and determine the corresponding bandwidth allocation for MTs with single-network and multi-homing calls;

M. Ismail and W. Zhuang, *Cooperative Networking in a Heterogeneous Wireless Medium*, SpringerBriefs in Computer Science, DOI: 10.1007/978-1-4614-7079-3_5,
© The Author(s) 2013

- The stochastic user mobility and call traffic models are necessary for designing the decentralized radio resource allocation mechanisms so as to investigate their associated impact on the system in terms of signalling overhead and processing time complexity;
- Concepts of short-term call traffic load prediction and network cooperation can help to reduce the amount of signalling overhead that is expected in a decentralized architecture. In addition, they allow for fast handover and hence support seamless service provision;
- There are two performance metrics for radio resource allocation, namely the amount of allocated resources per call and the corresponding call blocking probability. In this brief, we focus on the existing trade-off between these two metrics and present two design parameters that, when appropriately chosen, can strike a balance between the amount of allocated resources per call and the target call blocking probability;
- MTs should play an active role in the resource allocation operation, instead of being a passive service recipients in the networking environment. The mechanisms presented in this brief enable an MT with single-network service to select the best wireless access network available at its location and asks for its required bandwidth from that network. In addition, an MT with multi-homing service can determine the required bandwidth share from each available network so as to satisfy its total required bandwidth.

5.2 Future Research Directions

In this brief, we mainly focus on exploiting cooperative networking in a heterogeneous wireless access medium to enhance service quality to mobile users. Cooperative networking can also help to improve overall network performance. One research direction that is not well investigated in the context of cooperative networking is related to green radio communications [26]. This research direction is motivated by the increasing BS energy consumption of the wireless networks, which affects the annual profits of the service providers and has a significant impact on the environment due to the associated CO_2 emissions [11, 26, 38, 69]. Cooperative networking can help to improve the networks energy efficiency in the following ways.

Energy saving in wireless communication networks can be achieved at several levels. One level focuses on the layout of networks and their management that takes into account the changing call traffic load patterns along the day. This is referred to as dynamic planning [26]. Dynamic planning exploits the call traffic load fluctuations to save energy, by switching off some BSs when and where the traffic load is light. This is performed under the assumption that the radio coverage and service provision for the off cells can be taken care of by the remaining active BSs. However, this may result in coverage holes and/or inter-cell interference. These shortcomings can be avoided if dynamic planning is incorporated with network cooperation. Networks with overlapped coverage area can save energy by alternately switching on and off their radio resources according to call traffic load fluctuations [26]. The call traffic load then is carried on by the remaining active networks in the geographical region. Hence, it is required to develop a decentralized optimal resource (BSs/APs and radio transceivers) on-off switching policy that adapts to the fluctuations in the call traffic load and maximizes the amount of energy saving under service quality constraints.

In this brief, we focus on multi-homing services mainly to enhance the users QoS. It has been shown that, for a given network-MT pair, there exists an optimal transmission rate with minimal energy consumption [67]. However, this energy-optimal transmission rate may not provide the rate required by the MT. Hence, multi-homing radio resource allocation can be used to achieve energy saving through allocating the energy optimal transmission rate from each network to the MT, while satisfying the MT required total transmission rate. Cooperative networking concepts need to be developed in order to enable a decentralized implementation of the energy-efficient resource allocation mechanisms.

References

1. http://www.gams.com
2. Akyildiza, I., Lo, B., Balakrishnana, R.: Cooperative spectrum sensing in cognitive radio networks: a survey. Physical Communication **4**(1), 40–62 (2011)
3. Akylidiz, I.F., Xie, J., Mohanty, S.: A survey of mobility management in next-generation all-ip based wireless systems. IEEE Wirel. Commun. Mag. **11**(4), 16–28 (2004)
4. Alshamrani, A., Shen, X., Xie, L.: Qos provisioning for heterogeneous services in cooperative cognitive radio networks. IEEE J. Select. Areas Commun. **29**(4), 819–830 (2011)
5. Benameur, N., Fredj, S.B., Delcoigne, F., Oueslati-Boulahia, S., Roberts, J.W.: Integrated admission control for streaming and elastic traffic. In: Proceedings of 2nd International Workshop Quality of Future Internet Services, pp. 69–81 (2001)
6. Bertsekas, D.P.: Non-linear programming. Athena Scientific, Massachusetts (2003)
7. Bharati, S., Zhuang, W.: Cah-mac: cooperative adhoc mac for vehicular networks. IEEE J. Select. Areas Commun. (to appear)
8. Blau, I., Wunder, G., Karla, I., Sigle, R.: Decentralized utility maximization in heterogeneous multicell scenarios with interference limited and orthogonal air interfaces. EURASIP J. Wirel. Commun. Networking (2009)
9. Bonami, P., Kilinc, M., Linderoth, J.: Algorithms and software for convex mixed integer nonlinear programs. Technical Repeport 1664, Computer Sciences Department, University of Wisconsin-Madison (2009)
10. Bussieck, M.R., Vigerske, S.: Minlp solver software. In: Cochran, J.J. (ed.) Wiley Encyclopedia of Operations Research and, Management Science, vol. 1, pp. 114–125. Wiley, Hoboken (2011)
11. Cai, L.X., Yongkang, L., Tom, H.L., Shen, X., Mark, J.W., Poor, H.V.: Dimensioning network deployment and resource management in green mesh networks. IEEE Wirel. Commun. Mag. **18**(5), 58–65 (2011)
12. Cavalcanti, D., Agrawal, D.P., Cordeiro, C., Xie, B., Kumar, A.: Issues in integrating cellular networks, wlans, and manets: a futuristic heterogeneous wireless network. IEEE Wirel. Commun. Mag. **12**(3), 30–41 (2005)
13. Chebrolu, K., Rao, R.: On robust allocation policies in wireless heterogeneous networks. In: Proceedings of 1st International Conference on Quality of Service in Heterogeneous Wired/ Wireless Networks, pp. 198–205 (2004)
14. Chebrolu, K., Rao, R.: Bandwidth aggregation for real time applications in heterogeneous wireless networks. IEEE Trans. Mobile Comput. **5**(4), 388–402 (2006)
15. Chiang, M., Low, S., Calderbank, A., Doyle, J.: Layering as optimization decomposition: a mathematical theory of network architectures. Proc. IEEE **95**(1), 255–312 (2007)
16. Choi, H., Cho, D.: On the use of ad hoc cooperation for seamless vertical handoff and its performance evaluation. Mob. Netw. Appl. **15**(5), 750–766 (2010)

M. Ismail and W. Zhuang, *Cooperative Networking in a Heterogeneous Wireless Medium*, SpringerBriefs in Computer Science, DOI: 10.1007/978-1-4614-7079-3, © The Author(s) 2013

17. Fan, Y., Jiang, Y., Zhu, H., Shen, X.: Pie: cooperative peer-to-peer information exchange in network coding enabled wireless networks. IEEE J. Select. Areas Commun. **9**(3), 945–950 (2010)
18. Feldmann, A., Whitt, W.: Fitting mixtures of exponentials to longtail distributions to analyze network performance models. Perform. Eval. **31**(3–4), 245–279 (1998)
19. Ferris, M.C., Jain, R., Dirkse, S.: Gdxmrw: interfacing gams and matlab. Technical Report (2011)
20. Gerasenko, S.J., Rayaprolu, A., Ponnavaikko, S., Agrawal, D.K.: Beacon signals: what, why, how, and where? Computer **34**, 108–110 (2001)
21. Gross, D., Shortle, J.F., Thompson, J.M., Harris. C.M.: Fundamentals of Queueing Theory. Wiley series in probability and statistics. John Wiley & Sons, Hoboken (2008)
22. Grossmann, I.E.: Review of nonlinear mixed-integer and disjunctive programming techniques. Optim. Eng. **3**(3), 227–252 (2002)
23. Hong, D., Rappaport, S.S.: Traffic model and performance analysis for cellular mobile radio telephone systems with prioritized and nonprioritized handoff procedures. IEEE Trans. Veh. Technol. **35**(3), 77–92 (1986)
24. Ismail, M., Abdrabou, A., Zhuang, W.: Cooperative decentralized resource allocation in heterogeneous wireless access medium. IEEE Trans. Wireless Commun. (to appear)
25. Ismail, M., Zhuang, W.: A distributed resource allocation algorithm in heterogeneous wireless access medium. In: Proceedings of IEEE ICC (2011)
26. Ismail, M., Zhuang, W.: Network cooperation for energy saving in green radio communications. IEEE Wirel. Commun. Mag. **18**(5), 76–81 (2011)
27. Ismail, M., Zhuang, W.: Decentralized radio resource allocation for single-network and multi-homing services in cooperative heterogeneous wireless access medium. IEEE Trans. Wireless Commun. **11**(11), 4085–4095 (2012)
28. Ismail, M., Zhuang, W.: A distributed multi-service resource allocation algorithm in heterogeneous wireless access medium. IEEE J. Select. Areas Commun. **30**(2), 425–432 (2012)
29. Ismail, M., Zhuang, W., Yu, M.: Radio resource allocation for single-network and multi-homing services in heterogeneous wireless access medium. In: Proceedings of IEEE VTC Fall, pp. 1–5 (2012)
30. Koutsopoulos, L., Losifidis, G.: A framework for distributed bandwidth allocation in peer-to-peer networks. Perform. Eval. **67**(4), 285–298 (2010)
31. Laneman, J.N., Tse, D.N.C., Wornell, G.W.: Cooperative diversity in wireless networks: Efficient protocols and outage behavior. IEEE Trans. Inf. Theory **50**(12), 3062–3080 (2004)
32. Lasdon, L.S.: Optimization Theory for Large Systems Macmillan Series in, Operations Research. Macmillan, New York (1970)
33. Letaief, K., Zhang, W.: Cooperative communications for cognitive radio networks. Proc. IEEE **97**(5), 878–893 (2009)
34. Li, M., Claypool, M., Kinicki, R., Nichols, J.: Characteristics of streaming media stored on the web. IEEE/ACM Trans. Networking **5**(4), 601626 (2005)
35. Liang, H., Zhuang, W.: Cooperative data dissemination via roadside wlans. IEEE Commun. Mag. **50**(4), 68–74 (2012)
36. Liang, H., Zhuang, W.: Double-loop receiver-initiated mac for cooperative data dissemination via roadside wlans. IEEE Trans. Commun. **60**(9), 2644–2656 (2012)
37. Litjens, R., van den Berg, H., Boucherie, R.J.: Throughputs in processor sharing models for integrated stream and elastic traffic. Perform. Eval. **65**(2), 152–180 (2008)
38. Lu, R., Li, X., Liang, X., Lin, X., Shen, X.: Grs: the green, reliability, and security of emerging machine to machine communications. IEEE Commun. Mag. **49**(4), 28–35 (2011)
39. Luo, C., Ji, H., Li, Y.: Utility-based multi-service bandwidth allocation in the 4g heterogeneous wireless access networks. In: Proceedings of IEEE WCNC (2009)
40. Mandjes, M.R.H., Zuraniewski, P.: M/g/infinity transience, and its applications to overload detection. Perform. Eval. **68**(6), 507–527 (2011)

41. Mohanty, S., Akylidiz, I.F.: A cross-layer (layer 2 + 3) handoff management protocol for next generation wireless systems. IEEE Trans. Mobile Comput. **5**(10), 1347–1360 (2006)
42. Naghshineh, M., Acampora, A.: Qos provisioning in micro-cellular networks supporting multiple classes of traffic. Wireless Netw. **2**(3), 195–203 (1996)
43. Navarro, E.S., Lin, Y., Wong, W.S.: An mdp-based vertical handoff decision algorithm for heterogeneous wireless networks. IEEE Trans. Veh. Technol. **57**(2), 1243–1254 (2008)
44. Neumaier, A., Shcherbina, O., Huyer, W., Vinko, T.: A comparison of complete global optimization solvers. Math. Program. **103**, 335–356 (2005)
45. Niyato, D., Hossain, E.: Bandwidth allocation in 4g heterogeneous wireless access networks: a noncooperative game theoretical approach. In: Proceedings of IEEE GLOBECOM, pp. 4357–4362 (2006)
46. Niyato, D., Hossain, E.: A cooperative game framework for bandwidth allocation in 4g heterogeneous wireless networks. In: Proceedings of IEEE ICC, pp. 4357–4362 (2006)
47. Niyato, D., Hossain, E.: A noncooperative game-theoritic framework for radio resource management in 4g heterogeneous wireless access networks. IEEE Trans. Mobile Comput. **7**(3), 332–345 (2008)
48. Nosratinia, A., Hunter, T.E., Hedayat, A.: Cooperative communication in wireless networks. IEEE Commun. Mag. **42**(10), 74–80 (2004)
49. Palomar, D., Chiang, M.: A tutorial on decomposition methods for network utility maximization. IEEE J. Select. Areas Commun. **24**(8), 1439–1451 (2006)
50. Palomar, D., Chiang, M.: Alternative distributed algorithms for network utility maximization: framework and applications. IEEE Trans. Automatic Control **52**(12), 2254–2269 (2007)
51. Pei, X., Jiang, T., Qu, D., Zhu, G., Liu, J.: Radio-resource management and access-control mechanism based on a novel economic model in heterogeneous wireless networks. IEEE Trans. Veh. Technol. **59**(6), 3047–3056 (2010)
52. Piamrat, K., Viho, C., Ksentini, A., Bonnin, J.M.: Resource management in mobile heterogeneous networks: state of the art and challenges. Technical Report 6459, Institute national de recherche en informatique et en automatique (2008)
53. Sahinidis, N.V., Tawarmalani, M.: Baron: Gams solver manual. Technical Report (2011)
54. Schluter, M., Egea, J.A., Banga, J.R.: Extended ant colony optimization for non-convex mixed integer nonlinear programming. Comput. Oper. Res. **36**(7), 2217–2229 (2009)
55. Schluter, M., Gerdts, M., Ruckmann, J.J.: Midaco: new global optimization software for minlp. Technical Report (2011)
56. Shan, H., Zhuang, W., Wang, Z.: Distributed cooperative mac for multi-hop wireless networks. IEEE Commun. Mag. **47**(2), 126–133 (2009)
57. Shen, H., Basar, T.: Differentiated internet pricing using a hierarchical network game model. In: Proceedings of IEEE ACC, pp. 2322–2327 (2004)
58. Shen, W., Zeng, Q.: Resource management schemes for multiple traffic in integrated heterogeneous wireless and mobile networks. In: Proceedings of 17th International Conference on ICCCN, pp. 105–110 (2008)
59. Soleimanipour, M., Zhuang, W., Freeman, G.H.: Optimal resource management in wireless multimedia wideband cdma systems. IEEE Trans. Mobile Comput. **1**(2), 143–160 (2002)
60. Song, W., Cheng, Y., Zhuang, W.: Improving voice and data services in cellular/wlan integrated network by admission control. IEEE Trans. Wireless Commun. **6**(11), 4025–4037 (2007)
61. Song, W., Zhuang, W.: Resource allocation for conversational, streaming, and interactive services in cellular/wlan interworking. In: Proceedings of IEEE GLOBECOM, pp. 4785–4789 (2007)
62. Song, W., Zhuang, W.: Multi-service load sharing for resource management in the cellular/wlan integrated network. IEEE Trans. Wireless Commun. **8**(2), 725–735 (2009)
63. Song, W., Zhuang, W., Cheng, Y.: Load balancing for cellular/wlan integrated networks. IEEE Network **21**(1), 27–33 (2007)

64. Taha, A.M., Hassanein, H.S., Mouftah, H.T.: Max-min fairness based radio resource management in fourth generation heterogeneous networks. In: Proceedings of 9th International Symposium on Communication and Information Technology, pp. 208–213 (2009)
65. Tawarmalani, M., Sahinidis, N.V.: Global optimization of mixed integer nonlinear programs: a theoritical and computational study. Math. Program. 99(3), 563–591 (2004)
66. Truong, C., Geithner, T., Sivrikaya, F., Albayrak, S.: Network level cooperation for resource allocation in future wireless networks. In: Proceedings of the 1st IFIP Wireless Days (2008)
67. Wang, W., Wang, X., Nilsson, A.A.: Energy-efficient bandwidth allocation in wireless networks: algorithms, analysis, and simulations. IEEE Trans. Wireless Commun. 5(5), 1103–1114 (2006)
68. Zhang, Q., Jia, J., Zhang, J.: Cooperative relay to improve diversity in cognitive radio networks. IEEE Commun. Mag. 47(2), 111–117 (2009)
69. Zheng, Z., Cai, L.X., Zhang, R., Shen, X.: Rnp-sa: joint relay placement and sub-carrier allocation in wireless communication networks with sustainable energy. IEEE Trans. Wireless Commun. 11(10), 3818–3828 (2012)
70. Zhou, T., Sharif, H., Hempel, M., Mahasukhon, P., Wang, W., Ma, T.: A novel adaptive distributed cooperative relaying mac protocol for vehicular networks. IEEE J. Select. Areas Commun. 29(1), 72–82 (2011)
71. Zhuang, W., Ismail, M.: Cooperation in wireless communication networks. IEEE Wireless Commun. Mag. 19(2), 10–20 (2012)
72. Zhuang, W., Mohammadizadeh, N., Shen, X.: Multipath transmission for wireless internet access from an end-to-end transport layer perspective. J. Internet Technol. 3(1) (2012)